Waves

Fundamentals and dynamics

Series on wave phenomena in the physical sciences

Series Editor
Sanichiro Yoshida
Southeastern Louisiana University

About the series

The aim of this series is to discuss the science of various waves. It consists of several books, each covering a specific subject known as a wave phenomenon. Each book is designed to be self-contained so that the reader can understand the gist of the subject. From this viewpoint, the reader can read any book as a stand-alone article. However, it is beneficial to read multiple books as it would provide the reader with the opportunity to view the same aspect of wave dynamics from different angles.

The targeted readership is graduate students of the field and engineers whose background is similar but different from the subject. Throughout the series, it is intended to help students and engineers deepen their fundamental understanding of the subject as wave dynamics. An emphasis is laid on grasping the big picture of each subject without dealing with detailed formalism, and yet understanding the practical aspects of the subject. To this end, mathematical formulations are simplified as much as possible and applications to cutting edge research are included. The reader is encouraged to read books cited in each book for further details of the subject.

Other titles in this series

William Parkinson *What's the Matter with Waves? An Introduction to Techniques and Applications of Quantum Mechanics*

Wayne D Kimura *Electromagnetic Waves and Lasers*

David Feldbaum *Gravitational Waves*

Michail Todorov *Nonlinear Waves: Theory, Computer Simulation, Experiment*

Waves

Fundamentals and dynamics

Sanichiro Yoshida

Southeastern Louisiana University, USA

Morgan & Claypool Publishers

Rights & Permissions
To obtain permission to re-use copyrighted material from Morgan & Claypool Publishers, please contact info@morganclaypool.com.

ISBN   978-1-6817-4573-2 (ebook)
ISBN   978-1-6817-4572-5 (print)
ISBN   978-1-6817-4575-6 (mobi)

DOI   10.1088/978-1-6817-4573-2

Version: 20171201

IOP Concise Physics
ISSN 2053-2571 (online)
ISSN 2054-7307 (print)

A Morgan & Claypool publication as part of IOP Concise Physics
Published by Morgan & Claypool Publishers, 1210 Fifth Avenue, Suite 250, San Rafael, CA, 94901, USA

IOP Publishing, Temple Circus, Temple Way, Bristol BS1 6HG, UK

This book is dedicated to all my teachers at Azabu high school and Keio University, my parents and wife.

Contents

Preface

This book is part of a series entitled 'Wave Phenomena in the Physical Sciences'. When I was invited to edit a book series in engineering, the word 'waves' immediately came to my mind. There are two main reasons for this. First, waves are important in many fields of engineering. Mechanical engineers use ultrasound devices to inspect aircraft or find new sources of petroleum; electrical engineers design tracking systems using GPS (global positioning system) using the microwave signals from GPS satellites; medical engineers develop new heart monitoring systems using EKG (electrocardiogram) signals that read the heart beats based on the electric waves emitted by the heart. Second, waves often represent the underlying dynamics of observed phenomena at a fundamental level of physics. By definition, waves represent the spatiotemporal behavior of physical systems. Dynamics is a study of spatiotemporal characteristics exhibited by objects. It is not surprising that the wave characteristics constitute the essential part of some dynamics. Understanding the wave characteristics properly for a certain physical system often enables us to discover the beauty of the underlying physics that nature integrates into the system. For instance, by viewing the retarded potentials and associated electromagnetic waves as the messengers that charges use to inform other charges of their motion, we can understand relativistic electrodynamics.

Understanding waves and wave dynamics has been challenging to me. In high school physics we learned 'wave velocity = frequency × wave length' and I took it for granted. In college physics, we were taught 'a wave carries energy not the medium' and 'light is a wave but also a particle called the photon'. I somewhat embraced the first statement, but I did not understand it on a sound physical basis. The second statement was out of the question. The wave velocity formula from high school physics did not help me picture light as a particle. A number of references with beautiful illustrations of the wave–particle duality of light did not help me either. In graduate school, we were taught laser physics and quantum mechanics. We built lasers ourselves to use in our own research. Our lasers worked out well for our thesis work, but my hands-on experience did not help me understand why light can be a particle or why the Schrödinger equation has complex numbers. These were all mysteries to me.

It was not until recently that those mysteries started to make sense to me. After getting involved in various research projects such as gauge theory of material sciences and laser interferometric gravitational wave detection, I started to understand the physical meanings behind those mysteries. Through these experiences I have learned the importance of viewing a given concept from various angles and thereby digesting the concept, as opposed to knowing formulae and laws associated with the concept.

Thus, the goal of this book is to discuss various aspects of wave dynamics from as many perspectives as possible. Most of this book stemmed from my research logs and lecture notes. Because of this, the organization of this book may be different from other books on similar subjects. It consists of comments rather than a systematic explanation of concepts. Some readers may find that different levels of concepts are mixed in a given section, but I designed those sections in this manner because I believe it helps give a solid understanding of the concept.

Throughout the book I tried my best to break down each concept so as to be as easily digestible as possible, hoping that this facilitates grasping the core of the underlying physics. In some cases, I show the mathematical process to derive the physical formula in a step-by-step manner because often the understanding of those mathematical processes helps digest the physical meaning of the formula.

The target audience of this book is undergraduate students majoring in engineering science and graduate students majoring in engineering in general. I hope this book helps the readers understand other books in the series and develop new ideas in applications of wave dynamics to their own fields.

This book consists of four chapters. A summary of each chapter is as follows.

Chapter 1 discusses fundamentals of oscillations and waves, describing the big picture of wave dynamics as propagation of oscillation. The chapter starts off with general discussions of harmonic oscillation using a simple mechanical system. The equation of motion governing the harmonic oscillation and their solutions are discussed under various conditions. The dynamics is discussed both in the time and frequency domains. Towards the end, the chapter discusses how the equation of motion evolves to a wave equation.

Chapter 2 discusses mathematical aspects of waves. Wave equations are derived from the equation of motion for some simple cases and their solutions are discussed. Like chapter 1, wave dynamics are viewed in the time and frequency domains. The definitions of the amplitude, phase and velocity of waves along with their physical meanings are discussed in detail. Superposition of multiple waves and their behaviors are also discussed.

Chapter 3 discusses various properties of waves. Fundamentals of each property are explained with somewhat detailed mathematical formulations. I used as many as possible examples, using different types of waves. The readers can use this chapter as a reference. Some practical applications to advanced science and engineering are touched on.

Chapter 4 is a short description of the propagation of waves in association with some actual techniques. It is impossible to go into details of these techniques. I tried to include references for each concept. Interested readers are encouraged to read these references.

I would like to take this opportunity to express my sincere gratitude to a number of people. Through this book project I realized that I owe a great deal to all my high school and college teachers from whom I received excellent training. I am grateful to my parents who provided me with the opportunity to receive such an excellent education. I also thank my wife Yuko Yoshida for her constant support for my scientific activities. Finally, I would like to thank Paul Petralia of Morgan & Claypool Publishers LLC for his continuous support and patience.

Author biography

Sanichiro Yoshida

Sanichiro Yoshida received his undergraduate and graduate educations in the field of Electrical Engineering and Applied Physics from Keio University, Japan. During this period, he had the opportunity to work as a student researcher at the University of Colorado Boulder. Under the supervision of Dr. Author V. Phelps, he conducted research in gaseous electronics. From this experience, he learned basics of scientific research. After receiving a PhD in 1986 from Keio University, he has conducted experimental and theoretical research on various topics, including developments and applications of lasers, optical interferometry, optical and acoustical characterization of material strength, and deformation and fracture of solids. His recent research interest is focused on development of a wave theory of deformation and fracture of solids. Currently, he is a Professor of Physics at Southeastern Louisiana University where he enjoys research with students. When not doing science, he will be found on the Judo mats or tennis courts on campus.

Waves
Fundamentals and dynamics
Sanichiro Yoshida

Chapter 1

Introduction

1.1 Oscillation and wave

In soccer stadiums spectators often make a wave by raising their hands. Their motion looks like a wave because spectators raise their hands in a certain pattern. After the first spectator raises her hands, the second spectator to her right raises his hands with a certain time-delay. Subsequently, the third spectator to the right of the second spectator raises her hands with the same time-delay. In this fashion, the raising-hand motion propagates in the same direction. When the person to the left of the first spectator raises his hands, the wave completes the propagation for the entire stadium. The wave looks good when all the spectators raise their hands to around the same height with the same time-delay. If one of these components is missing, the hand-raising motion will still propagate but the wave will appear somewhat distorted. The first element is called the amplitude of the wave and the second is related to the phase of the wave.

A wave has both temporal and spatial periodicities. In the above example, the temporal periodicity is determined by each individual spectator. The first spectator who initiates the wave raises their hands with a certain period (the time their hand needs to complete one cycle of the waving action). The next person repeats the hand waving motion with the same period. In wave dynamics, the first spectator can be viewed as the source. The next person follows the same motion excited (driven) by the source. We can say that the frequency (the reciprocal of the period) is determined by the source.

The spatial periodicity, referred to as the wavelength, is determined by how fast the waving motion is transfered from one person to the next. How fast the waving motion is transferred is measured by the wave velocity. With the same frequency, the higher the wave velocity, the longer the wavelength. This relation is conveniently described by $v_p = \nu\lambda$ where v_p is the wave velocity, ν is the frequency, and λ is the wavelength. This type of the wave velocity is called the phase velocity.

It is interesting to note that while the first spectator (the source) can determine the frequency, they do not have control over the wave velocity. It totally depends on how quickly the next person copies the waving motion. In the case of an actual physical wave, this means that the medium decides the wave velocity. In other words, the wave velocity is a medium constant and the wavelength depends on the frequency determined by the source. As an example of an actual physical system, consider propagation of a sound wave through air and steel rails. It is well known that when a train approaches a station the sound made by the train cars can be heard earlier through the rail than air. In this context, the wheels of the train are the source of the sound and the two media, air and the steel, propagate the sound with different velocities.

Apparently, there must be some interaction between a given segment (spectator) and the next so that the oscillatory motion is transferred. It is natural to interpret this interaction as the force exerted by a segment to the next segment. From this viewpoint, we can say that the medium exerts a certain type of force that maintains the wave to propagate. In the case of waves on a string, as will be discussed shortly, the tension causes the transverse displacement of the string to alternate the direction around the neutral position. Acoustic oscillation is maintained by the elastic force of the medium. In these cases, the differential oscillatory motion along the spatial coordinate axis causes the oscillation to propagate as a wave. The mechanism of wave propagation in these cases is similar to the wave in the soccer stadium discussed above. In the case of electromagnetic waves, the synergetic interaction between the electric and magnetic fields via Farady's law and Ampère's law causes oscillatory behaviors of the two fields and their propagation as waves.

While the type of force or mechanism that maintains wave dynamics depends on the specific oscillatory dynamics and associated medium, there is one common factor; that is, the force is always toward the neutral position (equilibrium). This type of force, or more specifically the one whose magnitude is proportional to the displacement from the equilibrium and whose direction is opposite to the displacement, is known as the ELASTIC or spring force. Elastic force drives oscillatory dynamics that propagates as a wave in numerous physical systems. So, in the following section, we will start our discussion from the equation of motion due to elastic force. The solution to the equation of motion represents a harmonic oscillation. After discussing some properties of harmonic oscillations in the time domain, we will discuss the oscillation in the frequency domain. The frequency domain analysis is important not only because it is practical to see the system's response to the driving force but it also provides us with some insight that helps us understand the underlying physics. Finally, we will conclude this chapter by discussing a wave traveling through a string as an example of a physical system where a local oscillatory motion is transfered through the medium as a wave.

1.2 Oscillation due to elastic force

1.2.1 Elastic force

A spring and point mass system is one of the simplest forms that illustrate elastic (spring) force and related dynamics. We can put the spring force in the following form

$$f = -k_{sp}x. \tag{1.1}$$

Here f is the spring force vector, x is the displacement vector from the equilibrium, and the constant k_{sp} is referred to as the stiffness or the spring constant.

Consider that a point mass connected to a spring is being stretched in figure 1.1. As we all know well, we can calculate the potential energy by integrating the force over the path that the object is displaced. In the present case, we can evaluate the spring potential energy by calculating the work an external agent would need to displace the mass with no acceleration. The following integration provides us with the work, hence the spring potential energy U_{sp}

$$U_{sp}(x) = \int_0^x \boldsymbol{f}_{ex} \cdot d\boldsymbol{x} = \int_0^x (-\boldsymbol{f}_{sp}) \cdot d\boldsymbol{x}$$
$$= \int_0^x k_{sp} x \hat{x} \cdot dx \hat{x} = k_{sp} \int_0^x x \, dx = \frac{1}{2} k_{sp} x^2. \tag{1.2}$$

Here $\boldsymbol{f}_{sp}$ is the spring force exerted by the spring on the mass as described by equation (1.1), $\boldsymbol{f}_{ex}$ is the force exerted by an external agent on the mass against the spring force so that the mass is displaced for dx.

Although we derived equation (1.2) for a simple spring–mass system, the potential energy expression in this form provides us with a significant message for general cases; that is, whenever the potential curve is a quadratic function of the space coordinate variable, or can be approximated by a quadratic function, we can describe the force near the equilibrium as a spring force.

Consider a potential energy $U(x)$ expressed as a function of x as follows. Here x is the spatial coordinate variable that measures the displacement from the equilibrium point. By Taylor expanding $U(x)$, we obtain the following expression

$$U(x) = U(0) + U'(0)x + \frac{U''(0)}{2}x^2 + \frac{U^{(3)}(0)}{3!}x^3 + \cdots + \frac{U^{(n)}(0)}{n!}x^n + \cdots. \tag{1.3}$$

To find the force from equation (1.3), we simply differentiate it as follows

$$f(x) = -\frac{dU}{dx} = -U'(0) - U''(0)x - \frac{U^{(3)}(0)}{2}x^2 + \cdots. \tag{1.4}$$

The first term on the right-hand side of equation (1.4) is a constant force independent of the spatial variable, and therefore not important in oscillation dynamics. The second term represents the elastic force responsible for oscillation. We can interpret

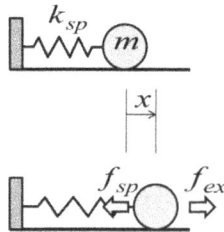

Figure 1.1. Point mass connected to a spring is being stretched by external force.

the coefficient U'' as the stiffness, or the spring constant k_{sp}. If we set $U'(0) = 0$ and neglect the second- and higher-order terms, equation (1.4) reduces to equation (1.1).

In dealing with actual physical systems, we often encounter cases where the force is directed to the equilibrium but it is not exactly proportional to the displacement. In these cases, we need to consider second- and higher-order terms in the force expression. Before concluding this section, let's consider up to the third term on the right-hand side of equation (1.4). We can deal with the situation by interpreting that the stiffness is a function of the spatial coordinate variable.

The third term represents the force that has a quadratic dependence on the spatial coordinate variable (the potential has a cubic dependence) [1]. By combining the first two terms of equation (1.4) and neglecting all the higher-order terms (without the negative sign in front), we can rewrite the right-hand side of this equation as follows

$$
\begin{aligned}
U''(0)x + \frac{U^{(3)}(0)}{2}x^2 &= \left(U''(0) + \frac{U^{(3)}(0)}{2}x\right)x \\
&= \left(k_{sp} + \frac{U^{(3)}(0)}{2}x\right)x \equiv k_{sp}(x)x.
\end{aligned}
\tag{1.5}
$$

Here, $k_{sp}(x) = k_{sp} + \frac{U^{(3)}(0)}{2}x$. In view of equation (1.5), we can interpret that the stiffness has a linear dependence on x, or the dynamics is nonlinear.

An excellent example of such a nonlinear elasticity is the inter-atomic force of solids. In most molecules, the potential curve is quadratic near the equilibrium. This means that the inter-atomic force near the equilibrium is a spring force, and that is the reason why most solids are linear elastic. However, if the displacement from the equilibrium is large enough to be outside of the quadratic part of the potential curve, the situation is different. On the short-distance side, as figure 1.2 (a) indicates, the potential has a steeper dependence on the inter-atomic distance, and consequently, the stiffness is higher than it is near the equilibrium. On the long-distance side, the potential has a less steep dependence on the inter-atomic distance, and therefore the stiffness is lower than near the equilibrium.

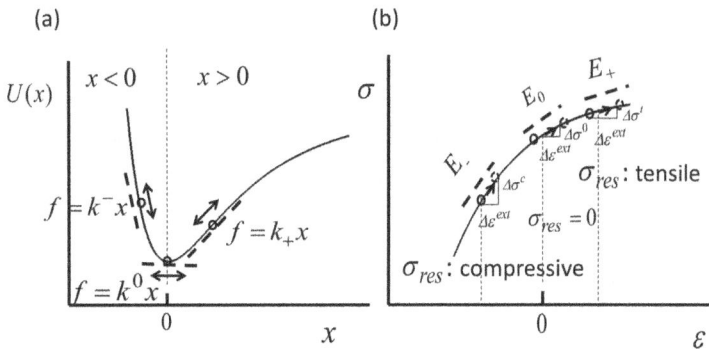

Figure 1.2. (a) Inter-atomic potential energy; (b) higher and lower elastic constants associated with compressive and tensile residual stresses.

These observations are applicable to residual stress analysis [2]. If a compressive residual stress locks a pair of neighboring atoms at a distance shorter than inter-atomic equilibrium distance, the corresponding elastic constant becomes higher than the nominal value. Figure 1.2 (b) indicates the situation. Conversely, if the residual stress is tensile, the elastic constant is lower than the nominal value. In these situations, the terms higher than the third order in equation (1.3) become significant. Acoustoelasticity exploits this fact in nondestructive evaluation of residual stresses [3]. When a part around a compressive residual stress is oscillated acoustically, the elevated elastic constant makes the acoustic velocity higher than the nominal value. If a part is around a tensile residual stress, the acoustic velocity is lower.

1.2.2 Equation of motion

Let's continue discussion of oscillation dynamics with a constant stiffness using the spring–mass system in figure 1.1. Under this condition, the only external force acting on the mass is the spring force. The corresponding equation of motion has the following form

$$m\boldsymbol{a} = m\frac{d^2\boldsymbol{\xi}}{dt^2} = -k_{sp}\boldsymbol{\xi}. \tag{1.6}$$

Here m is the mass connected to the spring whose spring constant is k_{sp}, $\boldsymbol{a}$ is the acceleration of m and $\boldsymbol{\xi}$ is the displacement vector from the equilibrium. Note that the acceleration and force are in line with the displacement.

When the system has a mechanism of energy dissipation in the form of velocity damping, we can add the following term on the right-hand side of equation (1.6) [4]

$$m\frac{d^2\boldsymbol{\xi}}{dt^2} = -k_{sp}\boldsymbol{\xi} - b\frac{d\boldsymbol{\xi}}{dt}. \tag{1.7}$$

Here b is referred to as the damping coefficient. As we will see in the next section, velocity damping causes the oscillatory system to decay exponentially. The actual decay rate is determined by the ratio of the damping coefficient to the mass. More details will be discussed soon.

When we align the direction of $\boldsymbol{\xi}$ to the x_s-axis, i.e., $\boldsymbol{\xi} = \xi\hat{x}_s$ where $\hat{x}_s$ is the unit vector, we can put equation (1.7) in the following form

$$m\frac{d^2\xi}{dt^2} + b\frac{d\xi}{dt} + k_{sp}\xi = 0. \tag{1.8}$$

Here the first term is the mass times acceleration, which originates in the left-hand side of equation of motion (1.6). The second and third terms represent the external forces acting on m where the first of these is the force proportional to the velocity (the first-order time derivative of the displacement) and the second is the elastic force exerted by the spring. The first force is known as the velocity damping force and it is an energy dissipative force that causes the oscillation to decay. The second force is an energy conservative force that is the source of oscillation. In the next section, we will discuss the energy conservative and energy dissipative dynamics using the same type of equation of motion as equation (1.8). The oscillatory and decaying behaviors of the mass will be discussed in detail.

In the next chapter, we will extend equation (1.8) to a wave equation. There, we will find that the energy conservative and dissipative properties of a medium correspond to the energy conservative and dissipative force terms in the equation of motion. These correspondences make sense if we interpret oscillation and wave dynamics as follows. In the oscillation dynamics, the spring acts as a massless entity that exerts external force on the mass. In the wave dynamics, each part of the medium acts as a spring exerting an external force on the neighboring parts. Therefore, the elastic property of the medium contributes to the oscillatory dynamics that keeps the wave propagating, whereas the energy dissipative property of the medium causes the wave to decay.

1.2.3 Oscillation as a solution to equation of motion

The solutions to differential equation (1.8) represent displacement of mass m from the equilibrium point where the spring force is null. When the right-hand side of a differential equation in this form is zero, we say that there is no source term for the differential equation. In the present context, since the term on the left-hand side has the dimension of force, no source term physically means that there is no external force on the mass except the spring force and the velocity damping force. In other words, there is no external agent that exerts force on the mass from outside of the spring–mass system. In this case, the displacement of the mass exhibits a decaying harmonic oscillation where the oscillation frequency is determined by the spring constant and mass. This is in contrast to the case of forced oscillation as we will discuss shortly. Since there is no external agent that forces the system to oscillate, this type of oscillation is referred to as an unforced oscillation.

If an external agent forces the mass to oscillate, the resultant oscillation is referred to as a forced oscillation. If the external force exerts a harmonic force (a force in the form of a sinusoidal function of time), the mass oscillates harmonically at the same frequency as the external force. Often this frequency is called the driving frequency. Unlike an unforced oscillation, a forced oscillation continues as long as the external agent is active in exerting the driving force. The damping in this case does not cause the oscillation to decay. Instead, it makes the energy transfer from the driving source to the mass less efficient. The oscillation of this type is referred to as a damped driven harmonic oscillation.

Mathematically, solving the equation of motion for an unforced oscillation is to find a homogeneous solution to the differential equation. On the other hand, solving the equation of motion for a forced oscillation is to find a general solution to the differential equation. A general solution consists of a homogeneous solution and a particular solution. In this section, we first discuss how to find a homogeneous solution. It will be shown that the homogeneous solution characterizes the spring system's oscillatory and damping behaviors depending on the relative magnitude of the spring constant and damping coefficient. Then we discuss the case when the mass is driven sinusoidally by an external agent.

Unforced oscillation. On describing an unforced oscillation, equation (1.8) represents the self-organized dynamics of a spring–mass system in response to an initial condition; if the mass is initially on the stretched side of the equilibrium, the spring force pulls it toward the equilibrium point. Due to the inertia, the mass passes the equilibrium point and stops at a compressed point where the kinetic energy is absorbed by the compressed spring energy. Subsequently, the compressed spring force pushes the mass back toward the equilibrium point, and the same pattern of motion continues. If the damping force is zero, this oscillation continues forever. If the damping force is finite, the oscillation continues until the total mechanical energy is dissipated by the damping mechanism.

For simplicity, let's rewrite differential equation (1.8) as follows

$$\frac{d^2\xi}{dt^2} + 2\beta\frac{d\xi}{dt} + \omega_0^2\xi = 0. \tag{1.9}$$

Here β and ω_0 defined by equations (1.10) and (1.11) are referred to as the decay constant and natural frequency of the oscillatory system.

$$\beta = \frac{b}{2m} \tag{1.10}$$

$$\omega_0 = \sqrt{\frac{k_{sp}}{m}} \tag{1.11}$$

In equation (1.10) b is the damping coefficient introduced in equation motion (1.7). As we discussed there, b causes the displacement to decay exponentially. The decay constant β represents the actual decay rate[1]. Equation (1.10) indicates that the decay rate is determined by the ratio b/m. We can intuitively understand that higher the damping coefficient or lighter the mass, the faster the system decays. A massive object is more difficult to change the state of motion than a light object, making the energy damping mechanism to take longer time to damp the oscillation. Equation (1.11) indicates that the oscillation frequency also depends on the mass for a given spring constant. Again, a massive object is harder to change the state of motion. Consequently, it takes longer time for the spring to switch the direction of oscillation at the turning point; hence the oscillation frequency is lower.

There are multiple ways to solve equation (1.9). Here we consider a way, which in my opinion, is mathematically the easiest and illustrates the nature of the dynamics most naturally. Take a look at equation (1.9). We find that the function ξ has the following property: if we multiply the constant ω_0^2 to the function itself, 2β to the first-order time-derivative of the function, and add the products to the second-order time derivative of the function, the answer is zero. This means that this function does not change its form through differentiations except for a constant multiplied to it. We know that the exponential function has such a property. So, we can use an exponential function of the following form as a test solution to equation (1.9)

---

[1] Some authors call $\beta = b/(2m)$ the damping coefficient while other authors call b/m the damping coefficient. In this book, the words "damping coefficient", "decay constant" are used intergengeably depending on the context to mean either $b/(2m)$ or b/m.

$$\xi(t) = \xi_0 e^{\lambda t}. \tag{1.12}$$

Substitution of solution (1.12) into equation (1.9) leads to the following equation

$$\lambda^2 \xi_0 e^{\lambda t} + 2\beta\lambda\xi_0 e^{\lambda t} + \omega_0^2 \xi_0 e^{\lambda t} = 0. \tag{1.13}$$

We immediately notice that we can divide equation (1.13) by ξ_0. This indicates that the constant ξ_0 can be any value to satisfy differential equation (1.9). In other words, ξ_0 does not determine the nature of the oscillation dynamics. Also, since equation (1.13) holds for any time t, we can assume $e^{\lambda t} \neq 0$ and therefore we can divide the equation by $e^{\lambda t}$ as well. Divisions of equation (1.13) by ξ_0 and $e^{\lambda t}$ result in the following quadratic equation of λ known as the characteristic equation for a second-order differential equation

$$\lambda^2 + 2\beta\lambda + \omega_0^2 = 0. \tag{1.14}$$

The two roots of equation (1.14) have the following form

$$\lambda_{\pm} = -\beta \pm \sqrt{\beta^2 - \omega_0^2}. \tag{1.15}$$

Equation (1.15) leads to the following general form for the solution to differential equation (1.9)

$$\xi(t) = \xi_{0+} e^{\lambda_+ t} + \xi_{0-} e^{\lambda_- t}, \tag{1.16}$$

where amplitude ξ_{0+} and ξ_{0-} can be determined by initial conditions in the displacement and velocity

$$\xi(0) = \xi_{0+} + \xi_{0-} \tag{1.17}$$

$$\dot{\xi}(0) = \xi_{0+}\lambda_+ + \xi_{0-}\lambda_- \tag{1.18}$$

Here $\cdot$ represents the differentiation in time. From equations (1.17) and (1.18), we can readily find ξ_{0+} and ξ_{0-} as follows

$$\xi_{0+} = \frac{\dot{\xi}(0) - \lambda_- \xi(0)}{\lambda_+ - \lambda_-} \tag{1.19}$$

$$\xi_{0-} = \frac{-\dot{\xi}(0) + \lambda_+ \xi(0)}{\lambda_+ - \lambda_-}. \tag{1.20}$$

The reason why we need initial conditions both in displacement and velocity is as follows. The original differential equation (1.9) is second-order in time. This means that we need to integrate the equation twice to find the function $\xi(t)$ as a solution. This process involves the use of two constants of integration; $\dot{\xi}(0)$ for the integration of $\ddot{\xi}$ and $\xi(0)$ for the integration of $\dot{\xi}$.

Now consider the behavior of the general solution to equation (1.9). In general, oscillatory motions represented by the equation of motion in the form of equation (1.9) can be classified into three types referred to as over damping, critical damping

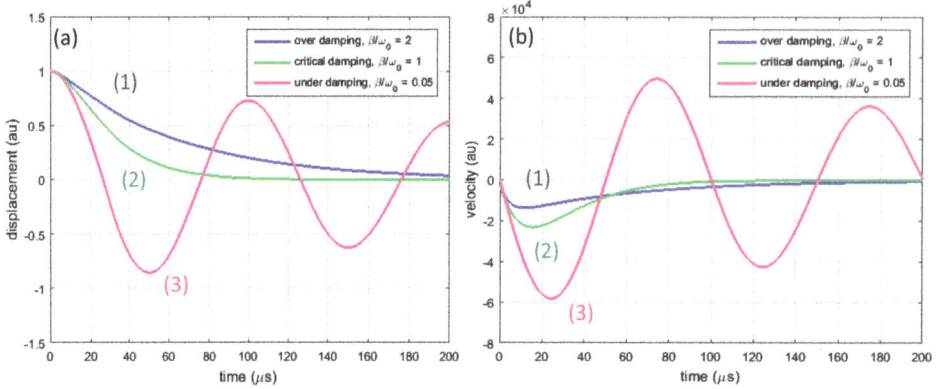

Figure 1.3. (a) Displacement solutions to equation (1.9) for (1) over damping, (2) critical damping and (3) under damping cases; (b) velocity solutions under the same conditions as displacement solutions.

and under damping. Depending on the degree of energy dissipation relative to energy conservation, the general solution to equation (1.9) can be any of these types. This relative energy dissipative nature of an oscillatory system is stipulated by the ratio β/ω_0. We can understand this through the following consideration. The parameter β indicates how fast the oscillation decays, whereas ω_0 indicates how fast the system repeats the oscillation. The former represents the energy dissipative and the latter the energy conservative nature of the system.

The ratio β/ω_0 has the following mathematical meaning; i.e., the specific equation (1.14) yields three types of root depending on this ratio. It yields two real roots if $\beta/\omega_0 > 1$; a real (double) root if $\beta/\omega_0 = 1$, and two complex roots if $\beta/\omega_0 < 1$. The three damping types have one-to-one correspondence to these three types of root, respectively. Below we discuss the behavior of the solution under the three damping conditions for initial conditions $\xi(0) = 1$ and $\dot{\xi}(0) = 0$. These initial conditions represent the situation where at $t = 0$ the mass is 1 m away from the equilibrium and at rest. Figure 1.3 illustrates these behaviors in the displacement and velocity.

(a) $\beta > \omega_0$ (over damping)

This condition is known as the over damping. The quantity inside the square root (the discriminant) of equation (1.15) is positive; hence the two roots λ are real numbers. Since $\left| \sqrt{\beta^2 - \omega_0^2} \right| < |\beta,|$ and $\beta > 0$, $\lambda_{+-} = -\beta \pm \sqrt{\beta^2 - \omega_0^2}$ are necessarily negative, real numbers. Hence, $\xi(t)$ expressed by equation (1.12) represents exponentially decaying displacement

$$\xi_{od}(t) = \xi_{od+}e^{-\gamma_1 t} + \xi_{od-}e^{-\gamma_2 t}. \tag{1.21}$$

Here I rewrite the decay constants with positive numbers as $\gamma_1 = -\lambda_+ > 0$ and $\gamma_2 = -\lambda_- > 0$ to emphasize that the solution (1.21) consists of two exponentially decaying terms. As $\gamma_1 < \gamma_2$, the second term decays faster than the first.

Let's consider a specific case with the initial conditions of $\xi(0) = 1$ and $\dot{\xi}(0) = 0$. Substituting $\xi(0) = 1$, $\dot{\xi}(0) = 0$ and $\lambda_+ - \lambda_- = 2\sqrt{\beta^2 - \omega_0^2}$ into equations (1.19) and (1.20), we obtain explicit expressions for ξ_{od+} and ξ_{od-} as follows

$$\xi_{od+} = \frac{1}{2} + \frac{\beta}{2\sqrt{\beta^2 - \omega_0^2}} \tag{1.22}$$

$$\xi_{od-} = \frac{1}{2} - \frac{\beta}{2\sqrt{\beta^2 - \omega_0^2}}. \tag{1.23}$$

Of course, substitution of equations (1.22) and (1.23) into equation (1.21) with $t = 0$ leads to $\xi(0) = 1$ and $\dot{\xi}(0) = 0$.

(b) $\beta = \omega_0$ (critical damping)

This condition is known as the critical damping. In this case, since the discriminant is null, the characteristic equation has only one root $\lambda = -\beta$. So we know that the solution to differential equation (1.9) decays with the decay constant β, which yields a solution in the form of $\xi(t) = \xi_0 exp(-\beta t)$. As we discussed above, however, this differential equation is second order in time. We need another term to satisfy both the initial condition for the displacement $\xi(0)$ and velocity $\dot{\xi}(0)$. It is natural to test the second term in the form of $f(t)exp(-\beta t)$. Thus we put the general solution as follows

$$\xi(t) = \xi_0 e^{-\beta t} + f(t)e^{-\beta t}. \tag{1.24}$$

Time derivatives of the second term on the right-hand side of equation (1.24), ξ_{cd2}, are as follows

$$\dot{\xi}_{cd2} = (\dot{f} - \beta f)e^{-\beta t} \tag{1.25}$$

$$\ddot{\xi}_{cd2} = (\ddot{f} - 2\beta\dot{f} + \beta^2 f)e^{-\beta t}. \tag{1.26}$$

Substituting equations (1.25) and (1.26) into differential equation (1.9) and dividing the resultant equation by the common nonzero factor $e^{-\beta t}$, we find as follows

$$(\ddot{f} - 2\beta\dot{f} + \beta^2 f) + 2\beta(\dot{f} - \beta f) + \omega_0^2 f = \ddot{f} + (\omega_0^2 - \beta^2)f = 0. \tag{1.27}$$

From the critical damping condition $\beta = \omega_0$, equation (1.27) leads to $\ddot{f} = 0$, i.e., $f(t)$ is a linear function of t. Thus we can write solution (1.24) as follows

$$\xi_{cd}(t) = \xi_{0cd1}e^{-\beta t} + \xi_{0cd2}te^{-\beta t}. \tag{1.28}$$

From initial conditions $\xi(0) = 1$ and $\dot{\xi}(0) = 0$, we find as follows

$$\xi_{cd}(0) = \xi_{0cd1} = 1 \tag{1.29}$$

$$\dot{\xi}_{cd}(0) = -\beta\xi_{0cd1} + \xi_{0cd2} = 0. \tag{1.30}$$

From equation (1.29) we find $\xi_{0cd1} = 1$ and substituting this into equation (1.30) we find $\xi_{0cd2} = \beta\xi_{0cd1} = \beta$. So, the explicit form of solution (1.28) for these initial conditions is as follows

$$\xi_{cd}(t) = (1 + \beta t)e^{-\beta t}. \tag{1.31}$$

(c) $\beta < \omega_0$ (under damping)

This case is called the under-damping or decaying oscillation, and is most important in oscillation dynamics. Since the discriminant is negative the two roots are complex numbers. Putting the square root of the discriminant as

$$\omega = \sqrt{\omega_0^2 - \beta^2} \tag{1.32}$$

we can express the two roots of the characteristic equation as follows

$$\lambda_{\pm} = -\beta \pm \sqrt{\beta^2 - \omega_0^2} = -\beta \pm i\sqrt{\omega_0^2 - \beta^2} = -\beta \pm i\omega. \tag{1.33}$$

Hence, we can write the solution to differential equation (1.9) in this case as follows

$$\begin{aligned}
\xi_{ud}(t) &= \xi_{0ud+}e^{-\beta t}e^{i\omega t} + \xi_{0ud-}e^{-\beta t}e^{-i\omega t} \\
&= \xi_{0ud+}e^{-\beta t}(\cos \omega t + i \sin \omega t) + \xi_{0ud-}e^{-\beta t}(\cos \omega t - i \sin \omega t) \\
&= e^{-\beta t}[(\xi_{0ud+} + \xi_{0ud-})\cos \omega t + i(\xi_{0ud+} - \xi_{0ud-})\sin \omega t] \\
&= \xi_{00}e^{-\beta t}\cos(\omega t - \delta_0).
\end{aligned} \tag{1.34}$$

where we used Euler's notation

$$e^{\pm i\theta} = \cos \theta \pm i \sin \theta \tag{1.35}$$

in the second line of equation, and used the following expressions for the amplitude and the constant part of the phase

$$\xi_{00} = 2\sqrt{\xi_{0ud+}\xi_{0ud-}} \tag{1.36}$$

$$\tan \delta_0 = \frac{i(\xi_{0ud+} - \xi_{0ud-})}{(\xi_{0ud+} + \xi_{0ud-})}. \tag{1.37}$$

ξ_{00} and δ_0 are determined by the initial condition.

Similar to the over damping and critical damping cases, we derive the explicit form of ξ_{ud} for the initial conditions $\xi_{ud}(0) = 1$ and $\dot{\xi}_{ud}(0) = 0$. From equation (1.33), in this case $\lambda_+ - \lambda_- = 2i\omega$. Therefore, from equations (1.19) and (1.20), we obtain the following expressions

$$\xi_{0ud+} = \frac{1}{2} + \frac{\beta}{2i\omega} = \frac{1}{2} - i\frac{\beta}{2\omega} \tag{1.38}$$

$$\xi_{0ud-} = \frac{1}{2} - \frac{\beta}{2i\omega} = \frac{1}{2} + i\frac{\beta}{2\omega}. \tag{1.39}$$

Substituting equations (1.38) into (1.36) and (1.37), we obtain the following explicit expressions:

$$\tan \delta_0 = \frac{(\xi_{0ud+} - \xi_{0ud-})}{(\xi_{0ud+} + \xi_{0ud-})} = \frac{i(-i\beta/\omega)}{1/2 + 1/2} = \frac{\beta}{\omega} \tag{1.40}$$

$$\xi_{00} = \sqrt{4\xi_{0ud+}\xi_{0ud-}} = \sqrt{1 + \left(\frac{\beta}{\omega}\right)^2} = \sqrt{1 + \tan \delta_0^2} = \frac{1}{\cos \delta_0}. \tag{1.41}$$

Figure 1.3 plots sample solutions to equation (1.9) for (a) displacement and (b) velocity. The natural frequency is fixed at 10 kHz, or $\omega_0 = 20\pi \times 10^3$ (rad s^{-1}), and the decay constant varies depending on the damping condition as indicated in the graphs. The over damping case shows a simply decaying feature. The under damping case indicates oscillatory behavior. The critical damping case also indicates the decaying feature where the decay is faster than the over damping case. We can interpret this behavior of the critical damping as being a mixture of the other two; the system tries to oscillate by returning towards the equilibrium position but fails to swing to the other side passing the equilibrium due to the damping effect.

Under the condition of under damping, the damping mechanism hinders the oscillatory behavior of the system. As we can easily imagine, the higher the damping coefficient, the less oscillatory the system becomes. The reduction in the oscillatory nature affects the shape and peak frequency of the Fourier spectrum, as will be discussed below.

Figure 1.4 shows the displacement of a sample under damped oscillations in a longer time scale than figure 1.3 for two decay constants, $\beta = 1000$ and 3000 1/s with natural frequency 10 kHz. The top graph is the displacement as a function of time

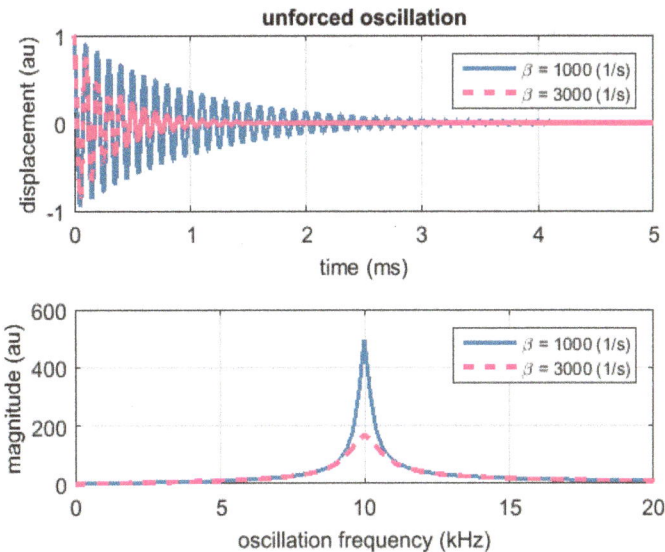

Figure 1.4. Behavior of unforced oscillation in time and frequency domains.

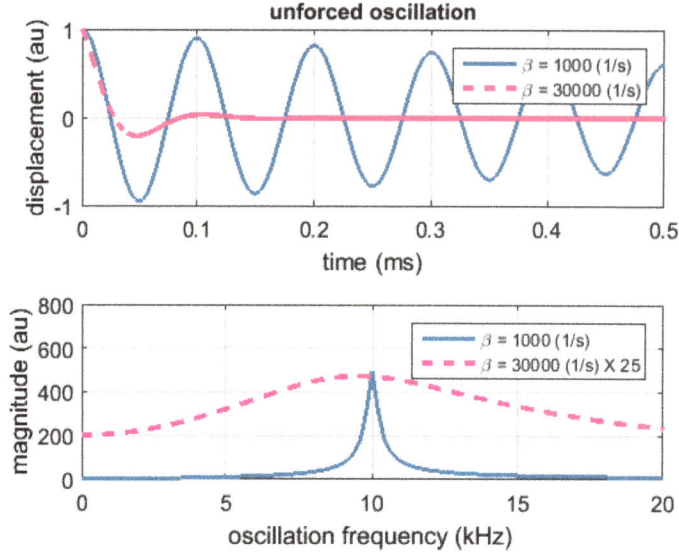

Figure 1.5. Behavior of unforced oscillation in time and frequency domains for higher damping case.

and the lower one is the Fourier power spectrum for each case. It is seen that with the increase in the damping effect, the displacement decays faster in time. In the frequency domain, the spectrum is broadened around the peak frequency, and at the same time, the peak height is lowered.

According to equation (1.32) the oscillation frequency decreases as the decay constant β increases. We can intuitively understand this tendency by considering the limit case of maximum damping as follows; if the damping effect keeps increasing, the system eventually loses its oscillatory behavior, which means that the frequency reduces to zero. In reality, however, the reduction in the oscillation frequency is unnoticeable unless the damping is very strong. Indeed, the lower plot of figure 1.4 hardly shows the difference in the peak frequency between the two damping cases. Figure 1.5 plots another case where the decay constant β is increased by an order of magnitude from the higher damping case in figure 1.4. The time-domain plot indicates that the decay constant is so high that the displacement barely shows oscillatory behavior for even one cycle. In the lower plot of this figure, the peak value of the spectrum for $\beta = 30\,000$ 1/s is multiplied by 25 for better visibility. Notice that the peak frequency is shifted to the lower frequency side by 0.5 kHz or so as compared with $\beta = 1000$ 1/s case. The spectral broadening is much greater than figure 1.4.

Forced oscillation. Differential equation (1.8) represents the case where there is no external agent exerting force on the oscillatory system consisting of a spring and mass; the mass in the system receives only the spring force and damping force. When an oscillatory system receives additional force exerted by an external agent, especially an alternating force oscillating at a frequency of Ω, the particular solution becomes more important. This type of oscillation is known as the damped driven harmonic oscillation (DDHO) [4], and seen in various physical systems such as alternating current circuits [5].

In this case, the corresponding equation of motion has a source term on the right-hand side of equation (1.8)

$$m\frac{d^2\xi}{dt^2} + b\frac{d\xi}{dt} + k_{sp}\xi = f_{ex}.$$ (1.42)

Here f_{ex} is the external force that forces the oscillatory system to behave in a certain way. Expressing the external force as $f_{ex}\,e^{i\Omega t}$, putting the particular solution in the form of

$$\xi = Ae^{i(\Omega t - \delta)}$$ (1.43)

and substituting it in equation (1.42), we find that

$$-mA\Omega^2 e^{i(\Omega t - \delta)} + ibA\Omega e^{i(\Omega t - \delta)} + k_{sp}Ae^{i(\Omega t - \delta)} = f_{ex}e^{i\Omega t}.$$ (1.44)

Dividing both hand sides of equation (1.44) by $me^{i\Omega t}$ and using equations (1.10) and (1.11), we obtain

$$A\Big[-\Omega^2 + i2\beta\Omega + \omega_0^2\Big]e^{-i\delta} = \frac{f_{ex}}{m}.$$ (1.45)

Further, by separating the terms of equation (1.45) into the real and imaginary parts, we obtain the following relations

$$A\Big[\big(\omega_0^2 - \Omega^2\big)\cos\delta + 2\beta\Omega\sin\delta\Big] = \frac{f_{ex}}{m}$$ (1.46)

$$\Big[2\beta\Omega\cos\delta - \big(\omega_0^2 - \Omega^2\big)\sin\delta\Big] = 0.$$ (1.47)

By solving equations (1.46) and (1.47), we find

$$\cos\delta = \frac{f_{ex}}{mA}\frac{\big(\omega_0^2 - \Omega^2\big)}{\big(\omega_0^2 - \Omega^2\big)^2 + (2\beta\Omega)^2}$$ (1.48)

$$\sin\delta = \frac{f_{ex}}{mA}\frac{2\beta\Omega}{\big(\omega_0^2 - \Omega^2\big)^2 + (2\beta\Omega)^2}.$$ (1.49)

Equations (1.48) and (1.49) lead to

$$\left(\frac{f_{ex}}{mA}\right)^2 \frac{\big(\omega_0^2 - \Omega^2\big)^2 + (2\beta\Omega)^2}{\Big[\big(\omega_0^2 - \Omega^2\big)^2 + (2\beta\Omega)^2\Big]^2} = \left(\frac{f_{ex}}{mA}\right)^2 \frac{1}{\big(\omega_0^2 - \Omega^2\big)^2 + (2\beta\Omega)^2} = 1.$$ (1.50)

From equation (1.50), we find the following expression for the oscillation amplitude

$$A = \frac{f_{ex}/m}{\sqrt{\big(\omega_0^2 - \Omega^2\big)^2 + (2\beta\Omega)^2}}.$$ (1.51)

As for the phase, from equations (1.48) and (1.49) we find

$$\tan \delta = \frac{2\beta\Omega}{\omega_0^2 - \Omega^2}. \tag{1.52}$$

Equations (1.51) and (1.52) tell us that when a harmonic oscillation system having natural (angular) frequency ω_0 and damping coefficient β is driven at (angular) frequency Ω, the amplitude and phase of the oscillation behave as follows.

(a) For a given damping coefficient, the amplitude is maximized when the driving frequency is equal to the natural frequency. This situation is referred to as resonance.

(b) As the damping coefficient increases, i.e., as the system becomes more energy dissipative, the amplitude decreases. Unlike the unforced (nondriven) case, the amplitude does not decay with time; instead it becomes a smaller value.

(c) The phase delay δ depends both on the damping coefficient and the driving frequency (Ω). Like the unforced oscillation case, the phase delay increases with the damping coefficient. When the driving frequency is equal to the resonant frequency ($\Omega = \omega_0$), the denominator of equation (1.52) becomes null, and consequently, $\tan \delta \to \infty$, i.e., $\delta = \pi/2$. As the driving frequency decreases from the resonant frequency ($\Omega < \omega_0$), $\tan \delta$ decreases towards 0 at the $\Omega = 0$ limit; hence δ decreases from $\pi/2$ towards 0. As Ω increases from ω_0 ($\Omega > \omega_0$), δ increases from $\pi/2$ towards π at the $\Omega \to \infty$ limit.

Figure 1.6 shows sample cases of forced oscillation with the same natural frequency (10 kHz) and two damping coefficients as figure 1.4 for three driving frequencies; $\Omega = 10.0$, 9.8 and 9.5 kHz. The above mentioned features (a)–(c) are observed as follows. (a) Of the three driving frequency cases, the case driven at the resonant frequency of 10 kHz shows the highest amplitude for either damping coefficient (decay constant β). (b) For each driving frequency case, the higher damping coefficient plot shows lower amplitude. In either case, the amplitude does not decrease with time. (c) When the driving frequency is equal to the resonant frequency, the displacement signal is sine function-like. Since the real part of the external force is a cosine function, this indicates that the phase delay is $\pi/2$. When the driving frequency is lower than the resonant frequency, the phase delay is less than $\pi/2$, i.e., the displacement signal is more cosine function-like than the case when the driving frequency is equal to the resonant frequency. This tendency increases as the driving frequency is reduced from 9.8 kHz to 9.5 kHz. With the driving frequency of 9.5 kHz, the displacement signal is almost cosine function-like for the smaller damping coefficient. Note that the higher damping coefficient case is less cosine function-like, reflecting the fact that for the same driving frequency the damping coefficient increases the phase delay.

Alternative way to derive amplitude and phase expressions. It is interesting to derive equations (1.51) and (1.52) in a slightly different way involving less mathematical manipulations. In this case, the phase delay is evaluated as part of the complex amplitude. We know that a particular solution has the same frequency as the driving

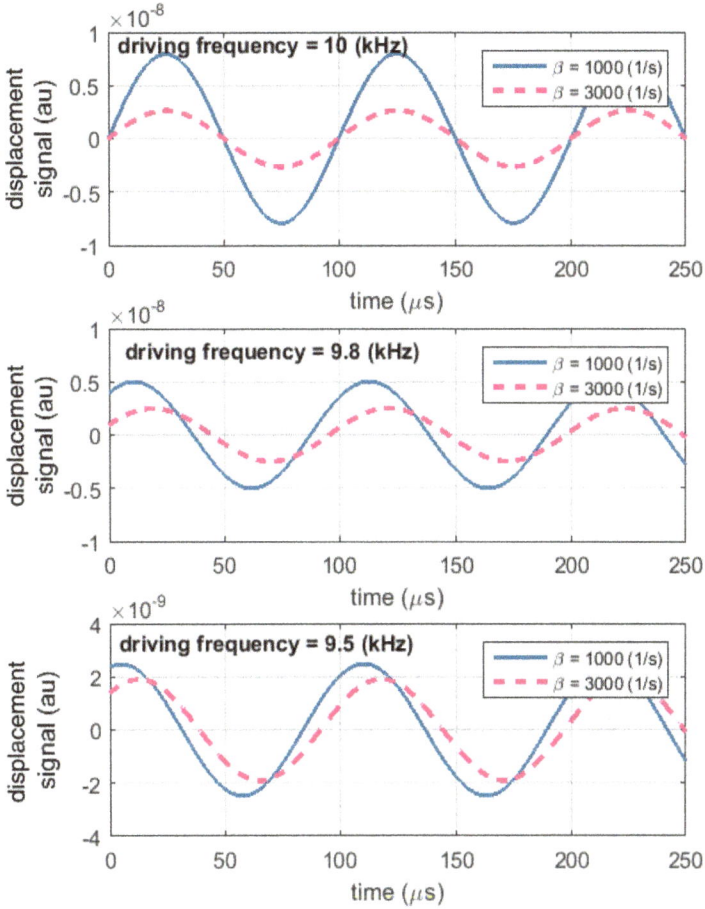

Figure 1.6. Behavior of forced oscillation in time domain.

frequency. So, we assume that the particular solution to differential equation (1.42) has the following form:

$$\xi(t) = \tilde{A}e^{i\Omega t} \qquad (1.53)$$

Here $\tilde{A}$ is the complex amplitude. Substituting equation (1.53) into differential equation (1.42) and using the same substitution of β and ω_0 as obtaining equation (1.45), we find as follows

$$\tilde{A} = \frac{f_{ex}/m}{\left(\omega_0^2 - \Omega^2\right) + i2\beta\Omega}$$

$$= \frac{f_{ex}}{m}\left[\frac{\omega_0^2 - \Omega^2}{\left(\omega_0^2 - \Omega^2\right)^2 + (2\beta\Omega)^2} - i\frac{2\beta\Omega}{\left(\omega_0^2 - \Omega^2\right)^2 + (2\beta\Omega)^2}\right]. \qquad (1.54)$$

Considering the absolute value and phase of $\tilde{A}$ in the expression equation (1.54), we can rewrite this complex number as follows:

$$\tilde{A} = \frac{f_{ex}/m}{\sqrt{\left(\omega_0^2 - \Omega^2\right)^2 + (2\beta\Omega)^2}} e^{i\phi} \tag{1.55}$$

where ϕ is given as:

$$\tan\phi = \frac{-2\beta\Omega}{\omega_0^2 - \Omega^2}. \tag{1.56}$$

From comparison of equations (1.52) and (1.56), we find the following relation

$$\phi = -\delta. \tag{1.57}$$

Substituting equation (1.57) into equation (1.55), and using the resultant expression in equation (1.53), we find

$$\xi(t) = \frac{f_{ex}/m}{\sqrt{\left(\omega_0^2 - \Omega^2\right)^2 + (2\beta\Omega)^2}} e^{i(\Omega t - \delta)}. \tag{1.58}$$

Equation (1.58) is identical to the solution we obtain by substituting equation (1.51) into equation (1.43).

1.3 Frequency domain analysis of oscillation

1.3.1 Fourier transform

The Fourier theorem states that a reasonably continuous periodic function can be expanded into a series of sine, cosine functions and their harmonics known as the Fourier series [6]. The Fourier transform is an extension of the Fourier series where the function has an infinite period. In the preceding section, we discussed that a solution to the equation of motion for an oscillatory system is fundamentally sine or cosine function-like. We observed that when an oscillatory system is unforced, it exhibits a characteristic frequency called the natural frequency, and that when the same oscillatory system is driven with a sinusoidal force, the solution exhibits a sinusoidal oscillation at the driving frequency.

A natural question is 'what if the same system is driven with a nonsinusoidal force?'. For instance, how does a spring–mass system behave if the mass is tapped. In this kind of situation, often Fourier transform is a powerful tool to analyze the system's response. Analysis based on Fourier transform is referred to as the frequency domain analysis. By Fourier transforming a nonsinusoidal driving force and analyzing the system's response in the frequency domain, we should be able to deal with the problem in the same fashion as a sinusoidal driving case. More specifically, we should be able to solve the equation of motion in the frequency domain. If we need to express the solution as a function of time, we can always inversely Fourier transform the frequency-domain solution.

The Fourier analysis is used in many fields of engineering and science. The relation between the external force and the system's response is referred to as the transfer function [7]; an indicator of how the action of the external agent (e.g. the external force on the mass) is transferred to the response of the system (e.g. the mass's displacement, velocity, etc). Mechanical engineers would apply the concept to designing a vibration isolation system for a comfortable passenger-car seat. In science, there is a specific field of study called spectroscopy [8]. In many situations, spectroscopy deals with the transfer function of a chemical system. When an analytical chemist identifies an unknown substance, he will apply a laser beam of various frequencies to the specimen and observe the resultant absorption. The substance absorbs the light when the frequency is at the resonance of the atomic transition. Naturally, the analysis is made in the frequency domain. The behavior of the atomic system can be described by the same type of equation of motion as (1.42), and the response can be interpreted as a transfer function. In the context of wave dynamics, this phenomenon can be interpreted as energy transfer from the light wave to the quantum mechanical wave function of the atomic system.

The Fourier transform and inverse Fourier transform are defined as follows

$$F(\omega) = \frac{1}{\sqrt{2\pi}} \int_{-\infty}^{\infty} f(t)e^{-i\omega t}\, dt \qquad (1.59)$$

$$f(t) = \frac{1}{\sqrt{2\pi}} \int_{-\infty}^{\infty} F(\omega)e^{i\omega t}\, d\omega. \qquad (1.60)$$

Here $i^2 = -1$.

To solve a differential equation in the frequency domain, we need to know how differentials behave in the Fourier domain. Consider Fourier transform of the first order derivative of function $f(t)$. Defining $g(t) \equiv \dot{f}(t)$, we can express the Fourier transform of $g(t)$ as follows

$$\mathcal{F}\{g(t)\} = \frac{1}{\sqrt{2\pi}} \int_{-\infty}^{\infty} \dot{f}(t)e^{-i\omega t}\, dt. \qquad (1.61)$$

Here $\mathcal{F}\{f(t)\}$ denotes the Fourier transform of function $f(t)$. Integrating the right-hand side by parts, we can rewrite equation (1.61) as follows

$$\begin{aligned}
\mathcal{F}\{\dot{f}(t)\} &= \frac{1}{\sqrt{2\pi}} [f(t)e^{-i\omega t}]_{-\infty}^{\infty} + \frac{i\omega}{\sqrt{2\pi}} \int_{-\infty}^{\infty} f(t)e^{-i\omega t}\, dt \\
&= \frac{i\omega}{\sqrt{2\pi}} \int_{-\infty}^{\infty} f(t)e^{-i\omega t}\, dt = i\omega F.
\end{aligned} \qquad (1.62)$$

Here the first term on the right-hand side vanishes at infinity because even if function $f(t)$ increases with t the exponential function approaches zero faster.

The above expression can be applied to higher order differentiations. By replacing $g(t)$ with $\dot{g}(t) = \ddot{f}(t)$ and expressing the Fourier transform of each function with the upper case letter, we obtain the following expression for the second-order derivative

$$\mathcal{F}\{\dot{g}(t)\} = i\omega G. \qquad (1.63)$$

From equation (1.62), $G = \mathcal{F}\{\dot{f}(t)\} = i\omega F$. Substituting this into equation (1.63), we find as follows

$$\mathcal{F}\{\ddot{f}(t)\} = \mathcal{F}\{\dot{g}(t)\} = i\omega G = i\omega(i\omega F) = (i\omega)^2 F. \tag{1.64}$$

In this fashion, we can express the Fourier transform of derivatives of the nth order in the form of $(i\omega)^n F$.

1.3.1.1 Solving equation of motion in frequency domain

Now we are in a position to express equation of motion (1.42) in the frequency domain. Applying the rule (1.64), we can rewrite equation (1.42) in the frequency domain as follows

$$ms^2\Xi(\omega) + bs\Xi(\omega) + k_{sp}\Xi(\omega) = (ms^2 + bs + k_{sp})\Xi(\omega) = F_{ex}(\omega). \tag{1.65}$$

Here $s = i\omega$ $(i^2 = -1)$, and $\Xi(\omega)$ and $F_{ex}(\omega)$ are the Fourier transform of $\xi(t)$ and $f_{ex}(t)$, respectively.

Transfer function. By solving equation (1.65) for $\Xi(\omega)$, we find

$$\Xi(\omega) = \frac{F_{ex}(\omega)}{ms^2 + bs + k_{sp}} \equiv F_{ex}(\omega)H(\omega). \tag{1.66}$$

Here $H(\omega)$ is referred to as the 'force-to-displacement' transfer function. When the explicit form of the Fourier transform of the input function F_{ex} is known, we can find the Fourier transform of the output $\Xi(\omega)$. Then using equation (1.60), we can find the output function $\xi(t)$ in the time domain. Figure 1.7 shows the transfer function of the above harmonic system represented by equation (1.42) with the natural frequency 10 kHz and damping coefficient 1000 and 3000 1/s. The top plot shows the magnitude of the transfer function and the bottom shows its phase.

Harmonic response. Now apply the concept of transfer function to the above example represented by equation (1.44) where we drove a harmonic oscillator with a sine or cosine function. According to equation (1.66), we can find the response of the harmonic system by multiplying the Fourier transform of the sinusoidal, driving function to the transfer function of the oscillatory system. More explicitly, the procedure is as follows. First, find the Fourier transform of the driving function, say $\cos(\Omega t)$. This is the input in the frequency domain

$$F_{ex}(\omega) = \frac{\delta(\omega - \Omega) + \delta(\omega + \Omega)}{2}. \tag{1.67}$$

Then, multiply $F_{ex}(\omega)$ to the transfer function $H(\omega)$ defined in equation (1.66). By inversely Fourier transforming the resultant function $\Xi(\omega)$, we can find the time-domain solution $\xi(t)$

$$\xi(t) = \frac{1}{\sqrt{2\pi}} \int_{-\infty}^{\infty} \left(\frac{\delta(\omega - \Omega) + \delta(\omega + \Omega)}{2(ms^2 + bs + k_{sp})}\right) e^{i\omega t} d\omega. \tag{1.68}$$

Figure 1.8 shows $\xi(t)$ obtained in this fashion for the same decay constant and driving frequencies as figure 1.6. As expected, the resultant waveforms of $\xi(t)$ are

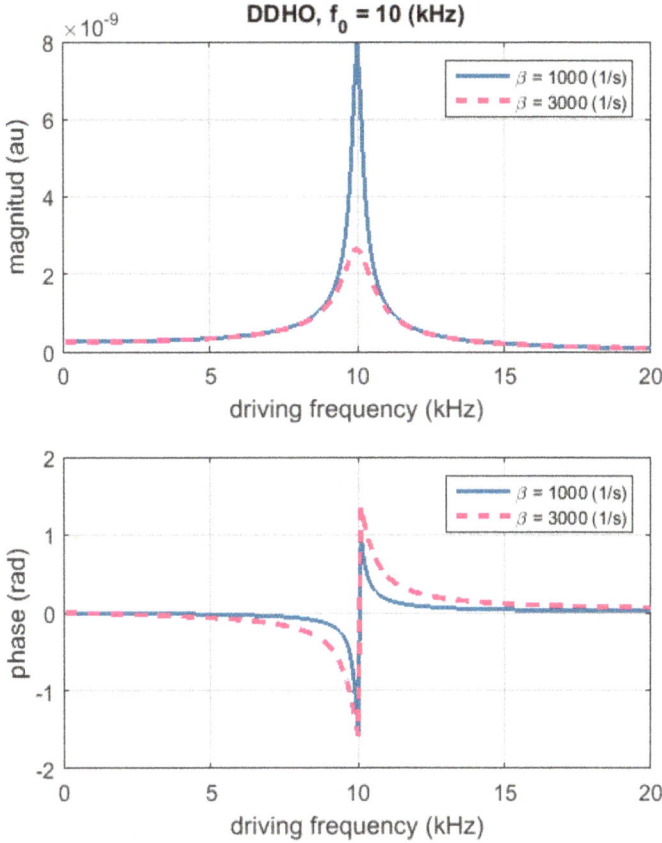

Figure 1.7. Transfer function of harmonic system represented by equation (1.42).

identical to figure 1.6. You may notice that the initial phases of figure 1.8 are different from 1.6. In the Fourier domain analysis, we consider the response of the system so the initial phase in the time domain is not important.

As a side issue, it is instructive to view the form of the Fourier transform of $\cos(\Omega t)$ in equation (1.67) in the following way. With the complex notation, $\cos \Omega t = (1/2)(\exp(i\Omega t) + \exp(-i\Omega t))$. The Fourier spectrum of $\exp(i\Omega t)$ is a sharp peak at $\omega = \Omega$. We can express this peak with a delta function as $\delta(\omega - \Omega)$. Similarly, we can express the Fourier spectrum of $\exp(-i\Omega t)$ as a sharp peak at $\omega = -\Omega$, i.e., $\delta(\omega + \Omega)$. The first term on the right-hand side of equation (1.67) represents the Fourier spectrum of $\exp(i\Omega t)$ and the second term that of $\exp(-i\Omega t)$.

Impulse response. We can express an impulse function occurring at time $t = 0$ with a delta function as $\delta(t)$. The Fourier transform of the delta function is a step function $E(\omega)$. By replacing $E(\omega)$ on the right-hand side of equation (1.66) with $E(\omega)$ and following the same procedures as the case of the harmonic response, we can find an impulse response. Figure 1.9 shows a time-domain solution found in this fashion for the same transfer function and decay constants as figure 1.8. The impulse excitation

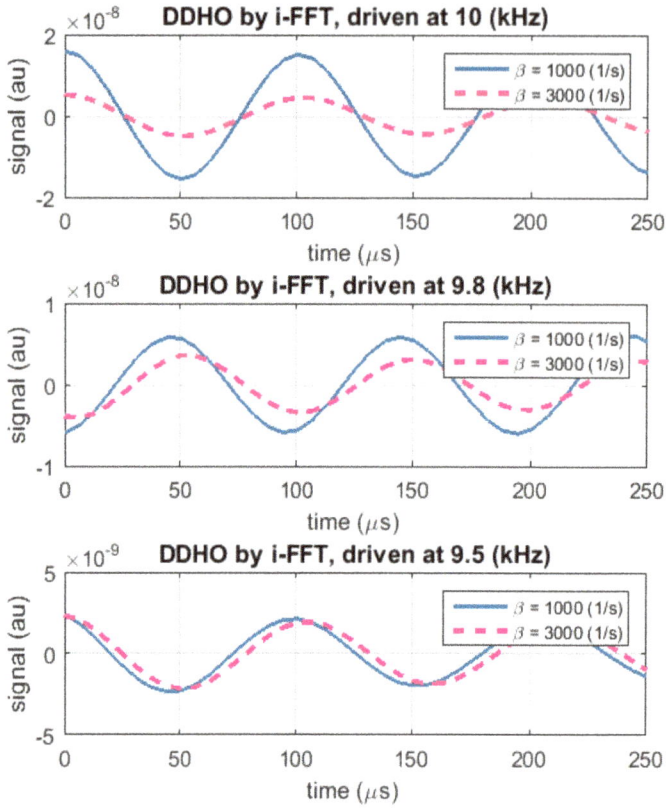

Figure 1.8. Time domain harmonic response found by inverse Fourier transform.

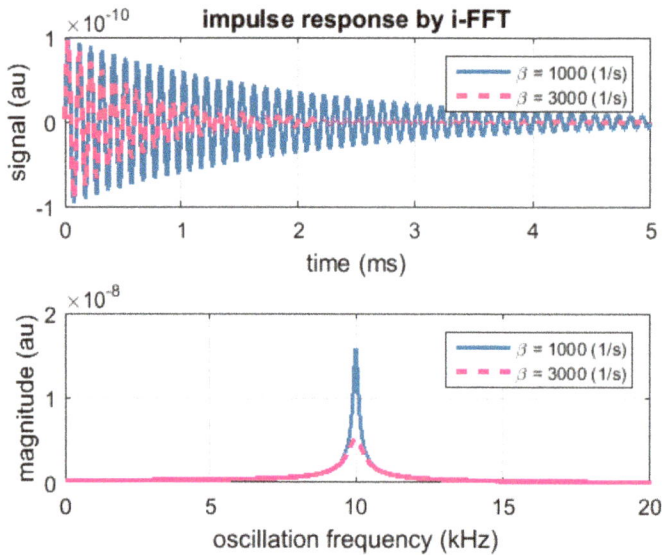

Figure 1.9. Time domain impulse response found by inverse Fourier transform.

of the harmonic oscillator corresponds to the unforced oscillation of the system. As expected, the solutions have the same temporal behavior as figure 1.4. The difference in the amplitude comes from the magnitude of the impulse input, which is arbitrary for the purpose of the discussion.

1.4 Wave

1.4.1 Oscillation to wave

Oscillation propagates as a wave when the neighboring section of the medium experiences the same or similar oscillation with a certain delay. In the discussion in the previous section, we derived oscillation from the equation of motion where the external force acting on the object is elastic. The argument assumes that the object is a point mass, or equivalently speaking, the elastic force is applied to the object uniformly. When we consider a finite volume for the object, the external force is not necessarily applied to the total mass uniformly. This is when the oscillatory motion becomes wave-like. We can view the medium as a series of infinitesimal segments connected to one another with a certain strength that each medium uniquely exhibits. The parameter characterizes this strength corresponding to the stiffness k_{sp}. Each segment acts as the external agent to exert force on the next segment. In this fashion, the type of the oscillatory motion discussed in the preceding section is transferred from one segment to the next.

Consider a simple case where oscillation propagates through a string. Figure 1.10 illustrates the external force acting on an infinitesimally short segment of a string. Call the external forces acting on the ends of the segment $T(x)$ and $T(x + dx)$. We can express the horizontal and vertical components of these forces as follows

$$T_x(x) = -|T(x)| \cos \theta_1 \tag{1.69}$$

$$T_x(x + dx) = |T(x + dx)| \cos \theta_2 \tag{1.70}$$

$$T_y(x) = -|T(x)| \sin \theta_1 \tag{1.71}$$

$$T_y(x + dx) = |T(x + dx)| \sin \theta_2. \tag{1.72}$$

Apparently, the segment does not move horizontally. Hence, the net horizontal force is null, and from equations (1.69) and (1.70) we obtain the following condition.

$$|T(x)| \cos \theta_1 = |T(x + dx)| \cos \theta_2. \tag{1.73}$$

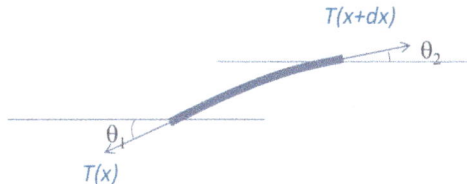

Figure 1.10. External force acting on two ends of a segment of an oscillating string.

When the oscillation amplitude is small, $\cos\theta_1 \cong \cos\theta_2$. This leads to the following conditions

$$|T(x)| = |T(x + dx)| = T. \tag{1.74}$$

Equation (1.74) indicates that the magnitude of the external force acting on each part of the string is constant over the entire length of the string. From this and equations (1.71) and (1.72), we can put the net vertical force acting on the segment of the string between x and $x + dx$ as follows

$$F_y^{net} = F_y(x + dx) - F_y(x) = T(\sin\theta_2 - \sin\theta_1). \tag{1.75}$$

Now consider relating angles θ_1 and θ_2 with the vertical displacement of the string at x, $\xi_y(x)$. Obviously,

$$\left.\frac{d\xi_y}{dx}\right|_x = \tan\theta_1 \cong \sin\theta_1 \tag{1.76}$$

$$\left.\frac{d\xi_y}{dx}\right|_{x+dx} = \tan\theta_2 \cong \sin\theta_2. \tag{1.77}$$

Here we use the small angle approximation $\sin\theta \cong \tan\theta$. From equations (1.75) (1.76) and (1.77), we find the net vertical force as follows

$$F_y^{net} = T\left(\left.\frac{d\xi_y}{dx}\right|_{x+dx} - \left.\frac{d\xi_y}{dx}\right|_x\right). \tag{1.78}$$

In the infinitesimal limit, we can express the inside of the parenthesis on the right-hand side of equation (1.78) using the secondary derivative of ξ_y with respect to x as follows

$$\left(\left.\frac{d\xi_y}{dx}\right|_{x+dx} - \left.\frac{d\xi_y}{dx}\right|_x\right) = \frac{d}{dx}\frac{d\xi_y}{dx}dx = \frac{d^2\xi_y}{dx^2}dx.$$

Hence, F_y^{net} becomes as follows

$$F_y^{net} = T\frac{d^2\xi_y}{dx^2}dx. \tag{1.79}$$

Thus, the equation of motion for this small segment of the string can be put in the following form

$$\left(\frac{m}{l}dx\right)\frac{d^2\xi_y}{dt^2} = T\frac{d^2\xi_y}{dx^2}dx. \tag{1.80}$$

Here m and l are the total mass and length of the string. By dividing both sides of equation (1.80) by dx, we obtain the following equation that describes the oscillation dynamics of each segment of the string that propagates as a transverse wave

$$\frac{d^2\xi_y}{dt^2} = \frac{Tl}{m}\frac{d^2\xi_y}{dx^2}. \tag{1.81}$$

The dimension of the term Tl/m appearing on the right-hand side is $[\text{N m kg}^{-1}] = [(\text{m s}^{-1})^2]$. This quantity is the square of the phase velocity. In this particular case, the phase velocity is a constant as T, l and m are all constants. As we will discuss in a later section, this fact leads to a nondispersive wave solution; i.e., the wave velocity does not depend on the frequency. In general, the wave velocity is a function of frequency by various mechanisms, which leads to the phenomenon known as dispersion.

It is easy to prove that a solution of the following form satisfies the wave equation (1.81)

$$\xi(t, x) = \xi_0 \sin (kx \pm \omega t). \tag{1.82}$$

Here ω and k are the angular frequency and wave number of the wave. The $\pm$ sign in the argument of the sine function indicates whether the wave travels in the positive or negative direction of the x-axis. (We will discuss the direction of wave in more detail in section 2.1.5). Equation (1.82) represents these forward- or backward-going waves comprehensively. Substitution (1.82) into the wave equation (1.81) leads to the following two expressions that represent the wave velocity for waves traveling either direction

$$v_p = \frac{\omega}{k} = \sqrt{\frac{Tl}{m}}. \tag{1.83}$$

Here ω/k literally represents the ratio of the temporal to spatial frequency, or defines the phase velocity of waves in the form of (1.82). On the other hand, $\sqrt{Tl/m}$ is the wave velocity resulting from the wave equation (1.81).

As briefly mentioned at the beginning of this chapter in the *waves in a soccer stadium* analogy, the physical system determines the wave (phase) velocity. The actual frequency and wave number of a wave on a string are determined by the boundary condition. In most cases our interest in string oscillation is the case when a standing wave is excited on the string. Therefore, here we discuss the boundary condition for a standing wave solution.

When waves travel in both directions with the same amplitude, a standing wave is formed. Adding the forward and backward traveling solutions (1.82) and using the mathematical identity regarding the addition of sine functions, we obtain the following expression of $\xi(t, x)$

$$\xi(t, x) = \xi_0(\sin (kx + \omega t) + \sin (kx - \omega t)) = 2\xi_0 \sin kx \cos \omega t. \tag{1.84}$$

Since the string is held at both ends ($x = 0$ and l), the value of ξ at both points must be always zero. Since $\sin(0) = 0$, the solution in the form of (1.84) automatically satisfies the boundary condition $\xi(0) = 0$. From the other boundary condition, we find the following conditions for k.

$$\sin(kl) = 0, \quad k_n = \frac{n\pi}{l}. \tag{1.85}$$

Here n is an integer referred to as the mode number. Equation (1.85) indicates that each integer n is associated with a wavelength λ_n. Since the wave velocity does not depend on the boundary conditions, there is a corresponding frequency for each λ_n. From equations (1.83) and (1.85), we find the following expressions for the wavelength and frequency of mode n

$$\lambda_n = \frac{2\pi}{k_n} = \frac{2l}{n} \tag{1.86}$$

$$\nu_n = \frac{v_p}{\lambda_n} = \frac{n\sqrt{\frac{Tl}{m}}}{2l} = \frac{n}{2}\sqrt{\frac{T}{ml}}. \tag{1.87}$$

The series of frequencies expressed by equation (1.87) are known as the characteristic frequency. It is apparent that addition of any of the modes satisfies the wave equation (1.81). In other words, under a given string and tension, the standing wave can have any combination of different modes with a certain weight. The overall wave can be expressed as a linear combination of different modes with a certain coefficient for each mode. The set of these coefficients is nothing but the Fourier spectrum. We will revisit this form of standing waves in the context of resonators in chapter 4 (section 4.2).

References

[1] Muir D D 2009 One-sided ultrasonic determination of third order elastic constants using angle-beam acoustoelasticity measurements *PhD Thesis* Georgia Institute of Technology Atlanta, GA, USA

[2] Yoshida S, Miural F, Sasaki T, Didie D and Rouhi S 2017 Optical analysis of residual stress with minimum invasion, *Society for Experimental Mechanics, 2017 Annual Meeting (12–16 June 2017 Indianapolis, USA)*

[3] Stobbe D M 2005 Acoustoelasticity in 7075-T651 Aluminum and dependence of third order elastic constants on fatigue damage *PhD Thesis* Georgia Institute of Technology, Atlanta, GA, USA

[4] Thornton S T and Marion J B 2004 *Oscillations, Classical Dynamics* 5th edn (Belmont, CA: Thomson) ch 3

[5] Tipler P A and Mosca G 2008 *Alternating-current circuits Physics for Scientists and Engineers* 6th edn (New York: Freeman) ch 29

[6] Boas M L 2006 *Mathematical Methods in the Physical Sciences* 3rd edn (London: Wiley) ch 7

[7] Boas M L 2006 *Mathematical Methods in the Physical Sciences* 3rd edn (London: Wiley) ch 8

[8] Diem M 2015 *Modern Vibrational Spectroscopy and Micro-Spectroscopy: Theory, Instrumentation, and Biomedical Applications* (Chichester: Wiley)

Chapter 2

Wave equations and solutions

2.1 Wave equation

The goal of this section is to derive wave equations from equations of motion, and discuss various features of wave equations.

2.1.1 Transverse waves

Refer to figure 2.1 and consider the equation of motion for the block at x of a continuous medium and associated wave dynamics. Call the vector components parallel to the x-axis p-components and those perpendicular s-components

$$\rho A \, dx \frac{d^2\xi}{dt^2} = df = \frac{df}{dx} dx. \tag{2.1}$$

Here df is the net external force acting on the block through the surfaces at $x = x$ and $x = x + dx$. In this case, since the external force acts on the block through the two surfaces, the net external force on the block is the differential force on the two surfaces. A is the area of the surface and ρ is the density. The force can be parallel or perpendicular to the boundary. First consider the parallel case. In this case, since the boundary is perpendicular to the x-axis, the force component responsible for the wave motion is the s-component, f_s. This type of force is referred to as shear force because it causes shear deformation [1]. Being an elastic force, $f_s(x)$ is proportional to the s-component of the stretch at x. Expressing this stretch component as the differential of ξ_s, we can write $f_s(x)$ in the following form

$$f_s(x) = k_s \frac{d\xi_s(x)}{dx} dx. \tag{2.2}$$

The equation of motion (2.1) becomes

$$\rho A \, dx \frac{d^2\xi_s}{dt^2} = \frac{df_s}{dx} dx = k_s \frac{d^2\xi_s(x)}{dx^2} (dx)^2. \tag{2.3}$$

doi:10.1088/978-1-6817-4573-2ch2
2-1

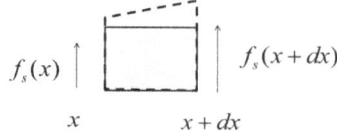

Figure 2.1. Generation of wave in a block of solid.

Here k_s is the stiffness of the medium in line with ξ_s. Defining the shear modulus as the elastic constant relating the shear stress σ_s (the shear force on a unit area of the surface) with the shear strain (the shear stretch per unit length) as follows

$$\sigma_s = \frac{f_s}{A} = G\frac{\partial \xi_s}{\partial x} \tag{2.4}$$

we find the following two expressions of f_s; one with the stiffness k_s and the other with shear modulus G

$$f_s = k_s \, d\xi = k_s\frac{\partial \xi_s}{\partial x}dx = GA\frac{\partial \xi_s}{\partial x}. \tag{2.5}$$

Comparing the two expressions connected by the rightmost equal sign in equation (2.5), we find the following relation between k_s and G

$$k_s \, dx = GA. \tag{2.6}$$

Using equation (2.6) and dividing both sides by $\rho A \, dx$, we can reformulate equation (2.3) as follows

$$\frac{d^2\xi_s}{dt^2} = \frac{k_s}{A\rho}dx\frac{d^2\xi_s(x)}{dx^2} = \frac{G}{\rho}\frac{d^2\xi_s(x)}{dx^2}. \tag{2.7}$$

Equation (2.7) is the equation for transverse waves. The word transverse here means that the wave propagates in a direction transverse to the oscillation; the s-component of the oscillating displacement vector travels as a wave along the x-axis.

The quantity G/ρ appearing on the right-hand side of equation (2.7) is worth paying attention to. We can view this quantity as the ratio of the secondary temporal derivative to the secondary spatial derivative of the displacement. Its unit is $[(1/s^2)/(1/m^2)] = [\mathrm{m^2\ s^{-2}}]$. Apparently, the dimension is velocity squared. The quantity $\sqrt{G/\rho}$ is known as the phase velocity of a wave, which we can view as a factor that converts the temporal periodicity of the displacement to the spatial periodicity. In the wave equation (2.7), this term consists of medium constants (G and ρ). This indicates that the wave propagates at a velocity determined by the medium.

2.1.2 Longitudinal waves

Repeating the same argument as above, we can derive the wave equation for a longitudinal wave; the case in which the force acting on the block is perpendicular to the boundary. All we need to do is to replace the s-component of the force and displacement with the p-component. Thus the equation of motion corresponding to (2.3) becomes as follows

$$\rho A \, dx \frac{d^2 \xi_p}{dt^2} = \frac{df_p}{dx} dx = k_p \frac{d^2 \xi_p(x)}{dx^2} (dx)^2. \tag{2.8}$$

Since we are interested in the wave traveling along the x-axis, x and dx remain the same as equation (2.3). The wave equation corresponding to (2.7) becomes

$$\frac{d^2 \xi_p}{dt^2} = \frac{k_p}{A\rho} dx \frac{d^2 \xi_p(x)}{dx^2} = \frac{E}{\rho} \frac{d^2 \xi_p(x)}{dx^2}. \tag{2.9}$$

Here k_p is the stiffness in line with ξ_p. The Young's modulus E is the longitudinal version of G, which satisfies

$$f_p = k_p \, d\xi_p = k_p \frac{\partial \xi_p}{\partial x} dx = EA \frac{\partial \xi_p}{\partial x}. \tag{2.10}$$

and therefore is related to k_p as follows

$$k_p \, dx = EA. \tag{2.11}$$

Equation (2.9) is referred to as the equation for a longitudinal wave. The wave is called longitudinal because unlike the case of transverse wave, the direction of oscillation is in line with the direction of propagation. As is the case of the transverse wave equation, the quantity E/ρ represents the square of the wave velocity, and consists of medium constants.

2.1.3 Plane wave equation

Using v_p to represent the phase velocity, we can comprehensively express equations (2.7) and (2.9) in the following form

$$\frac{\partial^2 \xi}{\partial t^2} = v_p^2 \frac{\partial^2 \xi}{\partial x^2}. \tag{2.12}$$

Here the phase velocity takes the form of the square root of the elastic constant divided by the density ρ

$$v_p = \sqrt{\frac{\kappa}{\rho}} \tag{2.13}$$

where κ represents the elastic constant in general. In the case of the transverse and longitudinal waves discussed above, $\kappa = G$ and $\kappa = E$, respectively. We will officially derive equation (2.13) later in this chapter (see equation (2.41) in sections 2.2.1).

Equation (2.12) indicates that the value of wave function ξ depends only on the coordinate variable along the axis of propagation. The variation in $\xi(x)$ along this axis comes from the variation in the phase. The amplitude remains constant as the wave travels. It also follows that the amplitude, hence the strength of the wave, is constant over a plane perpendicular to the axis of propagation. These types of waves are referred to as a plane wave.

The fact that equation (2.12) can comprehensively express transverse and longitudinal wave dynamics indicates that a solution to a wave equation in the form of equation (2.12) can describe both transverse and longitudinal waves. From the wave equation itself we cannot tell whether the solution is a transverse wave or longitudinal wave. It all depends on the restoring mechanism, as we observed in section 2.1. We need extra information to know if the wave is transverse or longitudinal.

For a three-dimensional wave, we can generalize the above wave equations as follows

$$\frac{\partial^2 \xi}{\partial t^2} = v_p{}^2 \nabla^2 \xi. \tag{2.14}$$

This generalization can be easily justified as follows. See figure 2.2.

$$d\xi = \frac{\partial \xi}{\partial s} ds. \tag{2.15}$$

By dividing both sides of equation (2.15) by ds and expressing the total differential of ξ in terms of partial derivatives with respect to x, y and z, we obtain the following expression for the strain ε as the spatial derivative of displacement ξ.

$$\begin{aligned}
\varepsilon &= \frac{d\xi}{ds} = \frac{\partial \xi}{\partial x}\frac{dx}{ds} + \frac{\partial \xi}{\partial y}\frac{dy}{ds} + \frac{\partial \xi}{\partial z}\frac{dz}{ds} \\
&= \frac{\partial \xi}{\partial x}\cos\theta_x + \frac{\partial \xi}{\partial y}\cos\theta_y + \frac{\partial \xi}{\partial z}\cos\theta_z = \frac{\partial \xi_x}{\partial x} + \frac{\partial \xi_y}{\partial y} + \frac{\partial \xi_z}{\partial z}.
\end{aligned} \tag{2.16}$$

Here $\cos\theta_x$ etc, are the directional cosines of the unit vector along the s-axis $\hat{s}$; $\hat{s} = \cos\theta_x \hat{x} + \cos\theta_y \hat{y} + \cos\theta_z \hat{z}$.

Replacing ξ with ϵ and repeating the same argument, we obtain the following expression for the secondary derivative of displacement

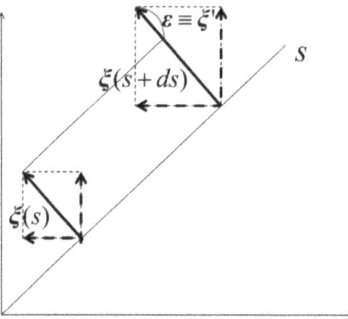

Figure 2.2. Change in displacement along s-axis.

$$\frac{d^2\xi}{ds^2} = \frac{d\varepsilon}{ds} = \frac{\partial\varepsilon_x}{\partial x} + \frac{\partial\varepsilon_y}{\partial y} + \frac{\partial\varepsilon_z}{\partial z}. \tag{2.17}$$

Since the strain vector's x, y and z components are spatial derivatives of displacement with respect to x, y and z respectively, we can replace $\partial\varepsilon_x/\partial x$ with $\partial^2\xi_x/\partial x^2$ (same for y and z). Thus,

$$\frac{d^2\xi}{ds^2} = \frac{\partial^2\xi_x}{\partial x^2} + \frac{\partial^2\xi_y}{\partial y^2} + \frac{\partial^2\xi_z}{\partial z^2} = \nabla^2\xi. \tag{2.18}$$

Equation (2.18) indicates that $\nabla^2\xi$ represents the secondary derivative of displacement along the ds-axis; i.e., it is indeed a one-dimensional wave whose spatial dependence is along the ds-axis. We call it a three-dimensional plane wave because the direction of the ds-axis is three-dimensional for a given Cartesian coordinate system. As is the case of wave equation (2.12), a wave equation in the form of equation (2.14) describes both transverse and longitudinal waves.

Wave as propagation of oscillation at natural frequency   It is interesting to express the phase velocity of a longitudinal wave in terms of the spring constant. For a longitudinal wave, we can use the Young's modulus for the elastic constant κ in the phase velocity expression (2.13). Using equation (2.11), we can write the phase velocity as follows

$$v_p = \sqrt{\frac{E}{\rho}} = \sqrt{\frac{k_{sp}\,dx}{\rho A}} = \sqrt{\frac{k_{sp}\,dx^2}{\rho(A\,dx)}} = \sqrt{\frac{k_{sp}}{m}}\,dx. \tag{2.19}$$

The term $\sqrt{k_{sp}/m}$ in equation (2.19) is the natural frequency of harmonic oscillation of mass m connected to a spring of stiffness k_p (equation (1.11)). Replacing $\sqrt{k_{sp}/m}$ with the natural frequency ω_0, we can rewrite equation (2.19) as follows

$$\omega_0 = \frac{v_p}{dx}. \tag{2.20}$$

We can view equation (2.20) as 'the wave velocity per unit length is the natural frequency of the unit length'. From this viewpoint, we can interpret the wave phenomenon as 'each unit length experiences harmonic oscillation at the natural frequency as a consequence of excitation (external force) by the neighboring unit'. This picture clearly exhibits the concept of a wave as continuous transfer of oscillatory motion from one unit to the next as we discussed in chapter 1. The fact that each unit oscillates at its natural frequency explains why 'the medium (the physical system) determines the wave velocity' as we discuss on several occasions in this book.

2.1.4 Decaying wave

So far we have discussed wave equations originating from equations of motion considering an elastic force as the only external force acting on the medium. In section 1.2.2 we discussed an equation of motion considering elastic force and velocity damping force as the external forces on a mass. To include a velocity damping mechanism to the dynamics, we need to add a term proportional to the first temporal derivative of the displacement on the right-hand side of equation (2.9) (or equation (2.7) for the transverse wave case). Using the decay constant β defined by equation (1.10), we can write the wave equation for this case as follows

$$\frac{d^2\xi}{dt^2} - \frac{E}{\rho}\frac{d^2\xi}{dz^2} + 2\beta\frac{d\xi}{dt} = 0. \tag{2.21}$$

Similarly to equation (2.14), we can generalize equation (2.21) to express a decaying wave propagating in a given direction as follows

$$\frac{d^2\xi}{dt^2} - \frac{E}{\rho}\nabla^2\xi + 2\beta\frac{d\xi}{dt} = 0. \tag{2.22}$$

Figure 2.3 compares the wave forms of nondecaying and decaying waves for several time steps. It is seen that the decaying wave not only shows a decrease in the peak values but also a slight delay in propagation. The latter represents the fact that when a wave decays, its velocity deviates from that of the nondecaying case as a function of frequency. This effect is known as dispersion of waves, which we will discuss in more detail later (section 3.2.2). Figure 2.4 is a close-up view of the two wave forms at $t = 1$ (s) in figure 2.3 in the range of $z = 0$ to $z = 3$. To emphasize the reduction in the wave velocity, the peak value at $z = 2$ for the decaying case is normalized to that of the nondecaying case. The slight delay in the decaying case is clearly seen.

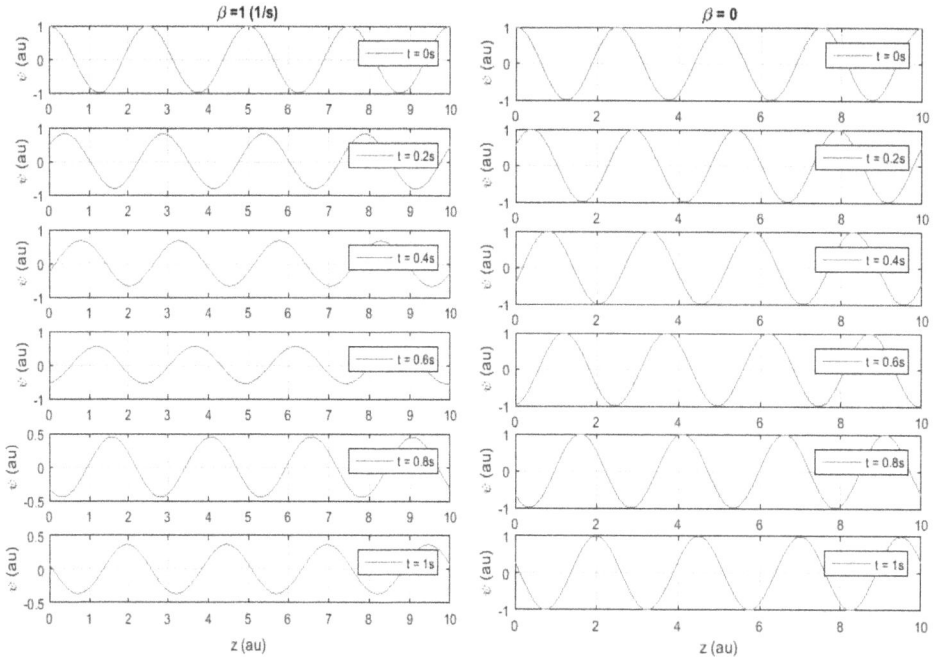

Figure 2.3. Slight decrease in wave velocity due to decay. Compare with nondecaying case on the right.

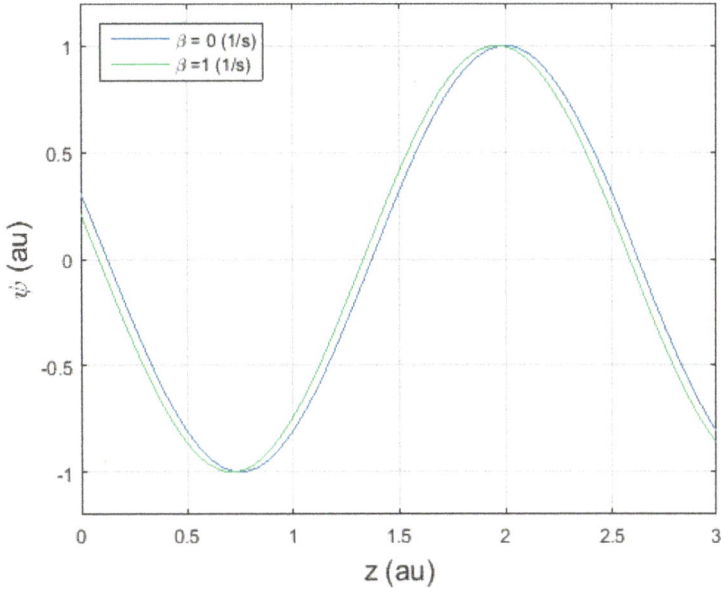

Figure 2.4. Phase delay due to decay observed near $z = 2$ peak.

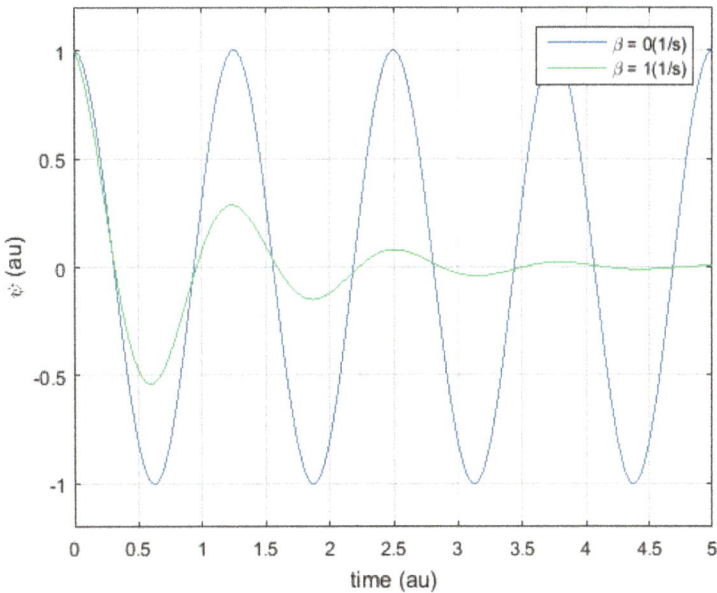

Figure 2.5. Time trend of decay wave shown in figure 2.3.

Figure 2.5 compares the decaying and nondecaying waves in figure 2.3 as a function of time. Notice that the large decay observed in the time plot contributes to the small phase delay. The reduction in the wave velocity due to a nonzero decay constant is small but certainly exists.

2.1.5 Some discussion on phase velocity

We discussed the concept of phase in chapter 1. The difference between an oscillation (temporal or spatial) and a wave is the way the phase changes. In an oscillation the phase varies as a function of time or space. In a wave, the phase varies as a function of time and space.

Consider the simplest case where the phase θ of a wave $\psi(\theta)$ is a linear function of t and z

$$\theta = \omega t \pm kz. \tag{2.23}$$

Here ω is the angular frequency in time and k is the angular frequency in space

$$\omega = \frac{\partial \theta}{\partial t} \tag{2.24}$$

$$k = \frac{\partial \theta}{\partial z}. \tag{2.25}$$

Since θ is a function of t and z, so is the wave function ψ

$$\psi(\theta) = \psi(t, z) = \psi(\omega t \pm kz). \tag{2.26}$$

Partially differentiating the function ψ with respect to time and space, we obtain the following set of equations.

$$\frac{\partial \psi}{\partial t} = \frac{\partial \psi}{\partial \theta}\frac{\partial \theta}{\partial t} = \omega \frac{\partial \psi}{\partial \theta} \tag{2.27}$$

$$\frac{\partial \psi}{\partial z} = \frac{\partial \psi}{\partial \theta}\frac{\partial \theta}{\partial z} = k \frac{\partial \psi}{\partial \theta}. \tag{2.28}$$

Here we used equations (2.24) and (2.25).

From equations (2.27) and (2.28) it follows that

$$\frac{\partial \psi}{\partial t} = \frac{\omega}{k}\frac{\partial \psi}{\partial z}. \tag{2.29}$$

Consider the meaning of the factor ω/k appearing on the right-hand side of equation (2.29). Obviously, this quantity represents the ratio of the temporal periodicity ω to the spatial periodicity k. Generally, this ratio is referred to as the phase velocity, v_p

$$v_p = \frac{\omega}{k}. \tag{2.30}$$

v_p is called the phase velocity because it literally represents the trace of a constant phase. See below for an explanation for this. From equation (2.23),

$$\frac{d\theta}{dt} = \frac{\partial \theta}{\partial t} \pm \frac{\partial \theta}{\partial z}\frac{dz}{dt}. \tag{2.31}$$

Express the condition of constant phase by setting $d\theta/dt = 0$. From equation (2.31), we find

$$\frac{\frac{\partial\theta}{\partial t}}{\frac{\partial\theta}{\partial z}} = \mp\frac{dz}{dt}. \tag{2.32}$$

From equations (2.24), (2.25), (2.30) and (2.32), we can express v_p as follows

$$v_p = \frac{\omega}{k} = \frac{\frac{\partial\theta}{\partial t}}{\frac{\partial\theta}{\partial z}} = \mp\frac{dz}{dt}. \tag{2.33}$$

Equation (2.33) explicitly indicates that v_p represents the change of the spatial coordinate for which the phase θ is constant over time; it represents the motion of a surfer who stays on the crest of the ocean wave, so to speak. Here the negative and positive signs in equation (2.33) correspond to those in equation (2.23). The former represents a wave traveling in the direction of the positive z-axis while the latter represents one traveling in the direction of the negative z-axis.

Repeating the same procedure as the one to obtain equation (2.29) and using equation (2.30), we can differentiate the wave function one more time in space and time to obtain the following wave equation

$$\frac{\partial^2\psi}{\partial t^2} = v_p^2\frac{\partial^2\psi}{\partial z^2}. \tag{2.34}$$

With the same mathematical procedure as the one to derive equation (2.14), we can generalize equation (2.34) to three-dimensional

$$\frac{\partial^2\psi}{\partial t^2} = v_p^2\nabla^2\psi. \tag{2.35}$$

The v_p^2 term appearing on the right-hand side of equations (2.34) and (2.35) is worth considering. This term results from differentiating the wave function in the form of equation (2.26) twice; each time we differentiate it with respect to time we multiply ω, and differentiate it with respect to space we multiply k. By differentiating the wave function twice, the $\mp$ in front of $\frac{dz}{dt}$ in equation (2.33) becomes positive. Consequently, the differential equation equation (2.34) or (2.35) represents both the forward-going wave $\psi(\omega t - kz)$ and the backward-going wave $\psi(\omega t + kz)$.

It is also worth noting that the secondary time-derivative in the wave equation originates from the equation of motion. Compare equation (2.34) with equation (1.81). We derived wave equation (1.81) from equation of motion (1.80) to consider the transverse wave traveling on a string. On the other hand, we derived equation (2.34) from the assumed wave solution (2.26) calling the ratio of temporal to spatial frequency the phase velocity ($v_p = \omega/k$). The comparison of these two equations leads to the following expression of the velocity for the transverse wave propagating on the string.

$$v_p = \frac{\omega}{k} = \sqrt{\frac{T}{m/l}}. \tag{2.36}$$

Equation (2.36) explicitly indicates that in this case the phase velocity is constant. Generally, the phase velocity ω/k is not necessarily a constant, causing the phenomenon known as dispersion of waves. This subject will be discussed later in section 3.2.

Note that what we did here is a generalization of what we had done at the end of chapter 1 (see the discussion near equation (1.83)). There we substituted a sine function into the string wave equation and obtained the phase velocity expression (2.36). Here we used a more general form (2.26) for the function to substitute into the wave equation and obtained the same phase velocity expression. This indicates that the solution to a wave equation in the form of (2.35) does not have to be a sinusoidal function.

2.2 Wave solutions

In this section, we discuss various solutions to wave equations (2.34) and (2.35). We start with the simplest case where the amplitude of the wave does not depend on the coordinates and the wave travels in line with one coordinate axis. This type of wave is referred to as a one-dimensional plane wave. The corresponding wave equation is equation (2.34).

2.2.1 Plane wave solutions

Rather than solving differential equation (2.34) with a certain mathematical technique, we assume a test wave function, substitute it into equation (2.34) and find the necessary conditions.

Apparently, a solution to differential equation (2.34) is a wave function. We can assume it being a function of the phase and use the phase in the form of equation (2.23). On the other hand, as we already discussed, a wave equation of this form is derived from an equation of motion representing dynamics due to elastic force. In chapter 1 we learned that a sinusoidal function is a natural solution to such an equation of motion. Combining these two factors, we can naturally use a test function in the following form

$$\psi(t, z) = \psi_0 \cos(\omega t \pm kz). \tag{2.37}$$

Now, we are using a cosine function in the test solution (2.37), but how about a sine function? Yes, of course a sine function works in exactly the same way. Then why don't we combine them? We know a powerful tool to combine them. Use the Euler's notation as we used in chapter 1 (equation (1.35))

$$e^{i\theta} = \cos\theta + i\sin\theta. \tag{2.38}$$

Applying equation (2.38), we can rewrite the test solution (2.37) as follows

$$e^{i(\omega t \pm kz)} = \cos(\omega t \pm kz) + i \sin(\omega t \pm kz). \tag{2.39}$$

Notice that the solution (2.37) constitutes the real part of the combined solution (2.39), and the sine function constitutes the imaginary part as the counterpart. From the complex theory [2], we know that a multiplication of the complex unit i to a complex number is equivalent to a phase delay of $\pi/2$[1]. So, the real part represents a solution $\pi/2$ ahead in phase to the imaginary part; indeed, a cosine function is $\pi/2$ ahead of the corresponding sine function, $\cos\theta = \sin(\theta + \pi/2)$.

Now we substitute the test solution (2.39) into the differential equation (2.34) using equation (2.13) for v_p, and obtain the following equation, known as the characteristic equation

$$\rho\omega^2 - \kappa k^2 = 0. \tag{2.40}$$

Equation (2.40) leads to the following expression of the wave (phase) velocity,

$$v_p = \frac{\omega}{k} = \sqrt{\frac{\kappa}{\rho}}. \tag{2.41}$$

Equation (2.41) explicitly indicates that the wave velocity propagating through this medium is determined by the elastic constant κ and density ρ.

From equations (2.37) and (2.40), the wave solution takes the following form

$$\psi(t, z) = \psi_0 \cos\left(\sqrt{\frac{\kappa}{\rho}}kt \pm kz\right). \tag{2.42}$$

Equation (2.42) represents a solution whose phase is zero at $t = z = 0$. Addition of a constant phase to the argument of the cosine function does not affect temporal or spatial differentiation of the function, meaning that it still satisfies the wave equation (2.34). Thus, we find the following expression as a solution to the equation for one-dimensional plane waves

$$\psi(t, z) = \psi_0 \cos\left(\sqrt{\frac{\kappa}{\rho}}kt \pm kz + \phi_0\right). \tag{2.43}$$

I would like to introduce the concept of wavefront at this point. The wavefront is a surface on which the phase of the wave is constant. The phase term in equation (2.43) is the argument of the cosine function. Since the spatial coordinate dependence of this phase term is limited to z, the constant phase surface is perpendicular to the z-axis. The wave number k measures the number of waves (the spatial periods) from a reference point in the unit of phase in rad along the axis of propagation. It is convenient to view the wave number k as representing the magnitude of a vector that carries the constant phase surface along the z-axis, or the axis of propagation. From this viewpoint, this vector is called the propagation vector. By definition, a

---

[1] Substituting $\pi/2$ to θ in equation (2.38) proves it.

propagation vector is perpendicular to the wavefront. In the case of three-dimensional wave, the propagation vector is in line with the s-axis (see equation (2.15)). If we express the propagation vector in the xyz-coordinate system, it has x, y and z components (see next section).

Extension to three-dimensions. We can find a solution to the general (three-dimensional) case of plane wave equation in the form of (2.14) by replacing kz with $\boldsymbol{k} \cdot \boldsymbol{r}$. Here $\boldsymbol{k}$ is the propagation vector and $\boldsymbol{r}$ is the coordinate vector

$$\boldsymbol{k} = k_x \hat{i} + k_y \hat{j} + k_z \hat{z} \tag{2.44}$$

$$\boldsymbol{r} = x \hat{i} + y \hat{j} + z \hat{z}. \tag{2.45}$$

Omitting the constant phase ϕ_0 for simplicity, we can express the three-dimensional solution as follows

$$\psi(t, z) = \psi_0 \cos\left(\sqrt{\frac{\kappa}{\rho}} kt \pm \boldsymbol{k} \cdot \boldsymbol{r}\right)$$

$$= \psi_0 \cos\left(\sqrt{\frac{\kappa}{\rho}} kt \pm (k_x x + k_y y + k_z z)\right). \tag{2.46}$$

Substitution of equation (2.46) into (2.14) leads to the following expression.

$$\frac{\kappa}{\rho} k^2 = v_p^2 \left(k_x^2 + k_y^2 + k_z^2\right). \tag{2.47}$$

By interpreting k_x, k_y and k_z as the x, y and z-component of $\boldsymbol{k}$ vector, we can replace the right-hand side of equation (2.47) with $|k|^2$. Since the left-hand side of equation (2.47) is equal to ω^2, we find that equation (2.47) reduces to equation (2.41). This indicates that the interpretation of the right-hand side of equation (2.46) being $|k|^2$ is appropriate, which in turn indicates that equation (2.46) is a solution to equation (2.14).

Decaying plane wave solution. The use of a test function to solve a wave equation works beautifully for the decaying case as well. In section 1.2.3, in order to allow the solution to represent the energy-dissipative dynamics, we multiplied the exponential decay term $\exp(-\beta t)$ to the sinusoidal term representing the oscillatory dynamics. We can use the same technique here. Using the exponential form (2.39) for the test solution, we can include the exponential decay term by making the angular frequency a complex number; $\omega = \omega_0 + i\omega_i$. We omit the constant phase ϕ_0 below as it does not change the gist of the discussion.

$$\xi = e^{i((\omega_0 + i\omega_i)t \pm kz)}. \tag{2.48}$$

Substituting equation (2.48) into the wave equation (2.21), we obtain the following characteristic equation.

$$\left(-\left(\omega_0^2 - \omega_i^2\right) + \frac{E}{\rho} k^2 - 2\beta\omega_i\right) + i(-2\omega_0\omega_i + 2\beta\omega_0) = 0. \tag{2.49}$$

Equation (2.49) leads to the following real and imaginary parts of the angular frequency

$$\omega_0 = \sqrt{\frac{E}{\rho}k^2 - \beta^2} \tag{2.50}$$

$$\omega_i = \beta. \tag{2.51}$$

With equations (2.50) and (2.51), the solution (2.48) becomes as follows

$$\xi = e^{-\beta t} \cos\left(\left(\frac{E}{\rho}k^2 - \beta^2\right)^{1/2} t \pm kz\right). \tag{2.52}$$

Solution (2.52) indicates that this wave decays with time.

Above, to make the wave solution to decay, we set the frequency to be a complex number. It is interesting to make the wave number complex. We know that as a wave, the spatial dependence of the phase should have a similar effect to the temporal dependence. Put the wave solution in the following form and examine how the solution behaves.

$$\xi = e^{i(\omega t \pm (k_0 - ik_i)z)}. \tag{2.53}$$

Substitution of equation (2.53) into equation (2.21) leads to the following characteristic equation

$$\left(-\omega^2 + \frac{E}{\rho}\left(k_0^2 - k_i^2\right)\right) + i\left(2\beta\omega - \frac{2E}{\rho}k_0 k_i\right) = 0. \tag{2.54}$$

By equating the real and imaginary parts of equation (2.54) to zero and after somewhat cumbersome algebraic manipulations, we find as follows

$$k_0 = \omega\sqrt{\frac{\rho}{2E}}\left[\sqrt{1 + \left(\frac{2\beta}{\omega}\right)^2} + 1\right]^{1/2} \tag{2.55}$$

$$k_i = \omega\sqrt{\frac{\rho}{2E}}\left[\sqrt{1 + \left(\frac{2\beta}{\omega}\right)^2} - 1\right]^{1/2}. \tag{2.56}$$

The real part of equation (2.53) takes the following form

$$\xi = e^{\pm ik_i z} \cos(\omega t \pm k_0 z). \tag{2.57}$$

With k_0 and k_i expressed by equations (2.55) and (2.56), equation (2.57) explicitly indicates that the wave decays as it travels.

2.2.2 Paraxial wave

In the preceding section, we considered plane waves. A plane wave has an infinite planar wavefront extending perpendicular to the axis of propagation, and its oscillation amplitude is the same everywhere on the planar wavefront. The propagation vector is perpendicular to the plane wavefront. These two situations indicate that the source of a plane wave is infinitely large, which is unrealistic[2].

More realistically, the amplitude, hence the energy, of a wave is concentrated near the axis of propagation. Naturally, the wavefront of a wave has the same size as the source at the origin. We can easily imagine that a point source generates a spherical wave, because the wave propagation must be spherically symmetric. Thus, a wave from a small source has a curved wavefront as figure 2.6 illustrates schematically. Since the propagation vector is perpendicular to the wavefront, the generated wave diverges as it travels. Waves whose energy is concentrated around the axis of propagation are called paraxial waves. Below, we find a paraxial wave solution to wave equation (2.35) and discuss its propagation behavior.

Helmholtz equation and paraxial approximation   We can find a paraxial wave solution by solving the spatial part of the wave equation. Consider variable separation in solving the wave equation (2.35). Put the solution in the following form[3]

$$\psi(t, r) = \psi_0 S(r) T(t).\tag{2.58}$$

Here $S(r)$ is a function of space but independent of time and $T(t)$ is a function of time but independent of space. Substitution of equation (2.58) into equation (2.35) and division of the resultant equation by ST yields the following

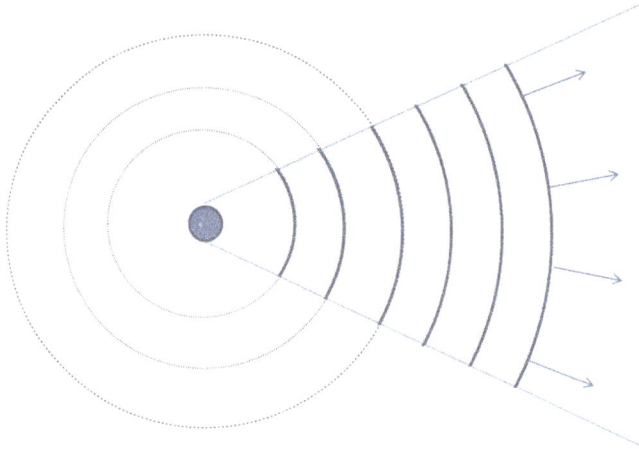

Figure 2.6. Diverging wave with a curved wavefront.

---

[2] Due to diffraction, a wave originating from a point source has a wavefront approximated as a plane at an infinite distance. Therefore, we can practically assume that the wavefront of a wave at a far distance from the source is plane. The light wave from the Sun is practically a plane wave on the Earth.
[3] See [4] for further details.

$$\frac{\ddot{T}}{T} = v_p^2 \frac{\nabla^2 S}{S} = -\omega^2. \tag{2.59}$$

Here the right-hand side of equation (2.59) comes from the assumption that the oscillation is in the form of

$$T(t) = \psi_0 e^{i\omega t}. \tag{2.60}$$

Substitution of equation (2.60) into equation (2.59) leads to the following differential equation for S known as the Helmholtz equation [3]

$$\nabla^2 S + k^2 S = 0. \tag{2.61}$$

where k is the magnitude of the propagation vector defined by equation (2.44)[4]. Note that since we are dealing with a wave propagation in the radial direction, k is constant in the normal direction to the wavefront at each point of propagation.

Assuming that $S(r)$ in equation (2.61) has exponential dependence along the propagation axis z, we can put $S(r)$ in the following form

$$S(r) = \psi(x, y, z)e^{-ikz}. \tag{2.62}$$

The exponential part of equation (2.62) indicates that the spatial part of the wave solution has a plane wavefront. Since this assumption is correct only near the axis of propagation, this assumption is referred to as the paraxial approximation. Yariv [4] explains paraxial approximation for light beams from a stable optical resonator. Schmerr [5] discusses paraxial ultrasonic waves.

For clarity, put the spatial differentiation in equation (2.61) in the following form

$$\nabla^2 = \nabla_t^2 + \frac{\partial^2}{\partial z^2}. \tag{2.63}$$

where the operator ∇_t^2 represents the transverse differentiation

$$\nabla_t^2 = \frac{\partial^2}{\partial r^2} + \frac{1}{r}\frac{\partial}{\partial r}. \tag{2.64}$$

Using equations (2.62), (2.63) and (2.64) in equation (2.61), we obtain the following differential equation

$$\nabla_t^2 \psi - 2ik\frac{\partial \psi}{\partial z} = 0. \tag{2.65}$$

In deriving equation (2.65), we ignore the second-order differentiation $\partial^2 \psi/\partial z^2$ assuming $\partial^2 \psi/\partial z^2 << \partial \psi/\partial z$.

Put the solution to equation (2.65) in the following form

$$\psi = \psi_0 e^{-i\left[P(z) + \frac{1}{2}Q(z)r^2\right]}. \tag{2.66}$$

---

[4] We assume a uniform medium here. If the medium is not uniform k can be a function of r; $k(r)$.

Substitution of equation (2.66) into equation (2.65) yields the following differential equation of $P(z)$ and $Q(z)$

$$-Q^2 r^2 - 2iQ - kr^2 \frac{\partial Q}{\partial z} - 2k \frac{\partial P}{\partial z} = 0. \qquad (2.67)$$

As equation (2.67) must hold for any r, it follows that

$$Q^2 + k \frac{\partial Q}{\partial z} = 0 \qquad (2.68)$$

$$\frac{\partial P}{\partial z} = -\frac{iQ}{k}. \qquad (2.69)$$

Thus, with the paraxial approximation, the wave equation governing the spatial part of the wave solution $S(r)$ reduces to the above pair of equations.

Equation (2.68) reads 'if you differentiate function $Q(z)$ with respect to z, the result is a negative constant times the function squared'. We all know that a function in the form of $1/z$ satisfies this property. So, let's put function Q in the following form

$$Q(z) = \frac{C}{Az + B}. \qquad (2.70)$$

Here A, B, C are constants to make the guessed function more general. Substitution of equation (2.70) into equation (2.68) indicates that $C = kA$. Thus,

$$Q(z) = k \frac{A}{Az + B}. \qquad (2.71)$$

It is convenient to define a function $q(z)$ as follows

$$q(z) = \frac{k}{Q(z)}. \qquad (2.72)$$

With $q(z)$ defined by (2.72), we can write equation (2.71) in the following form

$$q(z) = z + q_0. \qquad (2.73)$$

Here $q_0 = B/A$ is a constant.

Equation (2.73) tells us that the function $q(z)$ increases with z from the initial value q_0 at $z = 0$. From this standpoint, it is convenient to express equation (2.73) in the following form

$$q(z) = z + q(0). \qquad (2.74)$$

Equation (2.74) explicitly indicates that if we know the value of $q(0)$, we can evaluate $q(z)$ at any point on the z axis, meaning that we can find the solution $\psi(z)$ at a given z.

Gaussian beam.   With equations (2.72) and (2.73), we can write the solution (2.66) as follows

$$\psi = \psi_0 e^{-i\left[P(z)+\frac{1}{2}\frac{k}{z+q_0}r^2\right]} = \psi_0 e^{-iP(z)}e^{-i\frac{kr^2}{2(z+q_0)}}. \tag{2.75}$$

Here, consider the case when q_0 is a real number. In this case, the exponent of the second term on the right-hand side is imaginary, meaning that the absolute value of the second term is unity, $|e^{-i\frac{kr^2}{2(z+q_0)}}| = 1$. Hence, we can interpret this term as representing the phase of the wave. It follows that $\psi_0 e^{-iP(z)}$ represents the amplitude. However, this interpretation raises the following problem. $P(z)$ does not depend on r, and ψ_0 is a constant. If the $P(z)$ term represents the amplitude it is a constant with respect to r. In other words, although this solution represents a wave with a curved wavefront, its amplitude hence energy diverges infinitely. It does not represent a paraxial wave. So, we need to assume that q_0 is a complex number, and can put it as follows

$$q(z) = z + iz_0. \tag{2.76}$$

$q(z)$ is called the complex radius of the beam. Using equations (2.72) and (2.76), we can integrate equation (2.69) as follows

$$iP = \int_0^z \frac{1}{z+iz_0} = [\ln(z+iz_0]_0^z = \ln(z+iz_0) - \ln(iz_0) = \ln\left(1 - i\frac{z}{z_0}\right). \tag{2.77}$$

Now, expressing $1 - i\frac{z}{z_0}$ as a complex number in the form of $re^{-\theta}$ as

$$1 - i\frac{z}{z_0} = \sqrt{1+\left(\frac{z}{z_0}\right)^2}\, e^{-i\tan^{-1}\left(\frac{z}{z_0}\right)}, \tag{2.78}$$

we obtain the following equation

$$e^{-iP(z)} = e^{-\ln\left(\sqrt{1+\left(\frac{z}{z_0}\right)^2}e^{-i\tan^{-1}\left(\frac{z}{z_0}\right)}\right)} = \frac{1}{\sqrt{1+\left(\frac{z}{z_0}\right)^2}}e^{i\tan^{-1}\left(\frac{z}{z_0}\right)}. \tag{2.79}$$

The other term of equation (2.75) is

$$e^{-i\frac{kr^2}{2(z+q_0)}} = e^{-i\frac{kr^2}{2(z+iz_0)}} = e^{-\frac{izkr^2}{2(z^2+z_0^2)}}e^{-\frac{z_0kr^2}{2(z^2+z_0^2)}}$$
$$= e^{-i\frac{kr^2}{2z\left(1+\left(\frac{z_0}{z}\right)^2\right)}}e^{-\frac{k}{2z_0}\frac{r^2}{\left(1+\left(\frac{z}{z_0}\right)^2\right)}} \tag{2.80}$$
$$= e^{-i\frac{kr^2}{2R(z)}}e^{-\frac{1}{w_0^2}\frac{r^2}{\left(1+\left(\frac{z}{z_0}\right)^2\right)}}.$$

Here, w_0 and $R(z)$ are defined as

$$R(z) = z\left(1+\left(\frac{z_0}{z}\right)^2\right) = \frac{z^2+z_0^2}{z} \tag{2.81}$$

$$w_0 = \sqrt{\frac{2z_0}{k}} = \sqrt{\frac{\lambda z_0}{\pi}}. \tag{2.82}$$

We can view $R(z)$ as the coefficient to the quadratic function of r in the last line of equation (2.80), and can interpret it as the radius of curvature of the wavefront. Equation (2.81) indicates that the radius of curvature takes the minimum value of $2z_0$ at $z = z_0$ and infinity at $z = 0$.

Further, by introducing the following quantity,

$$w(z) = w_0 \sqrt{\left(1 + \left(\frac{z}{z_0}\right)^2\right)}. \tag{2.83}$$

and using equations (2.79) and (2.80), we can express equation (2.75) as follows

$$\psi(r, z) = \psi_0 \frac{w_0}{w(z)} e^{-\frac{r^2}{w(z)^2}} e^{i \tan^{-1}\left(\frac{z}{z_0}\right)} e^{-i\frac{kr^2}{2R(z)}}. \tag{2.84}$$

Equation (2.84) represents how the amplitude of the spatial part of the wave $S(r)$ defined by equation (2.62) varies as a function of the axial distance z (from the origin $z = 0$) and the radial distance r at each z. Although this is the amplitude for $S(r)$, it contains phase terms. We take a moment here to discuss the meaning of each term on the right-hand side of this equation.

The second and third terms $w_0/w(z)$ and $exp(-r^2/w(z)^2)$ are real, and represent the change in the amplitude of $S(r)$ as a function of z and r. From equation (2.83), we find that w_0 is the minimum value of $w(z)$ at $z = 0$; it is called the beam waist size. The term $w_0/w(z)$ represents the effect that the beam radius increases from the beam waist as z changes from $z = 0$. As equation (2.83) indicates, $w(z)$ increases both in the positive and negative directions of z.

We can relate the quantity z_0 we introduced in equation (2.76) to the beam waist size. Solving equation (2.82) for z_0,

$$z_0 = \frac{\pi w_0^2}{\lambda}. \tag{2.85}$$

z_0 is known as the Rayleigh length. From equation (2.76), we can interpret the Rayleigh length as the magnitude of the complex radius at $z = 0$; $q(0) = iz_0$. Equation (2.81) indicates that the minimum radius of curvature is twice that of the Rayleigh length.

The term $exp(-r^2/w(z)^2)$ represents the fact that the energy is concentrated near the z-axis, indicating that at a given z the amplitude decreases exponentially with r. In fact, this term takes the form of a Gaussian distribution. Thus, we call the wave in the form of equation (2.84) a Gaussian beam. At $r = w(z)$, the amplitude reduces to $1/e$ of the axial value (at $r = 0$). $w(z)$ is called the spot size of the beam. Equation (2.83) indicates that the spot size increases with z, meaning that the beam size increases as the wave travels away from the waist location ($z = 0$). It also tells us that the spot size at the Rayleigh length is square root two times the waist size ($w(z_0) = \sqrt{2} w_0$).

The other terms on the right-hand side of equation (2.84) represent the phase of the wave. The first of these (the one having the arc tangent function) indicates the advancement of the phase on the beam axis. The phase $\tan \phi = z/z_0$ is known as the Gouy phase. Using equation (2.85), we can express the Gouy phase in terms of the beam waist

$$\phi_G = \tan^{-1}\left(\frac{z}{z_0}\right) = \tan^{-1}\left(\frac{\lambda z}{\pi \omega_0^2}\right). \tag{2.86}$$

Equation (2.86) indicates that the smaller the Rayleigh length (z_0) or the beam waist size, the greater the Gouy phase advancement.

The second phase term with the radius of curvature $R(z)$ indicates that the phase varies quadratically on a plane perpendicular to the beam axis. The rate of the phase variation depends on the radius of curvature $R(z)$; the smaller the value of R, the faster the change in the radial phase. The r^2 in the exponent of this term indicates that a constant phase is a quadratic function of r, i.e., the wavefront is parabolic. In other words, a Gaussian wave has a spherical wavefront, whose radius of curvature changes according to equation (2.81).

Notice that equation (2.81) indicates that the smaller the Rayleigh length, hence the beam waist size, the faster the radius of curvature increases. We can understand this intuitively in figure 2.6, which illustrates the spherical wave fronts from a small source. If the source is smaller the radius of curvature at the source is smaller. Since the radius of curvature increases in proportion to the distance from the source, it takes a smaller increment in distance to double the radius of curvature. For a larger source size, the distance to double the radius of curvature is greater.

With equations (2.84) and (2.86), we can express the spatial part of the wave $S(r)$ defined by equation (2.62) in the following form.

$$S(r) = \psi(r, z)e^{-ikz} = \psi_0 \frac{w_0}{w(z)} e^{-\frac{r^2}{w(z)^2}} e^{-i(kz - \phi_G)} e^{-i\frac{kr^2}{2R(z)}}. \tag{2.87}$$

Together with equations (2.81) and (2.83), equation (2.87) describes the diverging behavior of a Gaussian beam completely.

Normally, the wavelength and beam waist size are given. So, it is practical to express the Rayleigh length with the wavelength and waist size using equation (2.85), and express the spot size using the Rayleigh length. Thus, from equation (2.83),

$$w(z)^2 = w_0^2 \frac{z^2 + z_0^2}{z_0^2} = \frac{\lambda}{\pi} \frac{z^2 + z_0^2}{z_0}. \tag{2.88}$$

From equations (2.76), (2.81) and (2.88),

$$\frac{1}{q} = \frac{1}{z + iz_0} = \frac{z - iz_0}{z^2 + z_0^2} = \frac{1}{R} - i\frac{\lambda}{\pi w(z)^2}. \tag{2.89}$$

We can use equation (2.89) to find the beam spot size and radius of curvature from $q(z)$, which we can find at a given z using equation (2.76).

2.2.3 Amplitude and phase of waves

The amplitude and phase fully define a sinusoidal wave. According to Fourier theorem, periodic functions are generally expanded into sine and cosine functions. Thus, we can say that the amplitude and phase define waves in general.

We can discuss the amplitude and phase of a sinusoidal wave most conveniently with the use of Euler's notation (2.38). Consider a wave function $\psi_0 \cos(\omega t \pm kz)$ on a complex plane, figure 2.7. Here the amplitude ψ_0 is the radius and the phase $\theta = \omega t \pm kz$ is the angle from the real axis. With the passage of time, the phase advances with the rate ω. In figure 2.7, this corresponds to counterclockwise movement of the tip of the radius with angular velocity of ω. If the wave decays, the radius decreases with the passage of time. If the wave grows, the radius increases. As long as the frequency remains the same, the angular velocity of the tip of the radius is the same.

Amplitude.   The amplitude of a wave is directly related to the energy carried by the wave. The physical dimension of energy is work, the product of force and displacement. As will be discussed in detail in section 3.1.2, we can characterize a physical wave with two quantities, one associated with the force and the other displacement. Both quantities travel together as waves having the same phase velocity and frequency (hence the same wavelength). We call the former the force-like quantity and the latter the displacement-like (velocity-like) quantity. (In oscillation dynamics, displacement times angular frequency is velocity. Thus displacement and velocity are proportional to each other as long as the frequency of the wave is constant. The product of displacement and force waves represents wave energy and the product of velocity and force waves represents wave power.) A sound wave carries the acoustic power via a pair of pressure (force-like) and velocity (velocity-like) waves. Here,

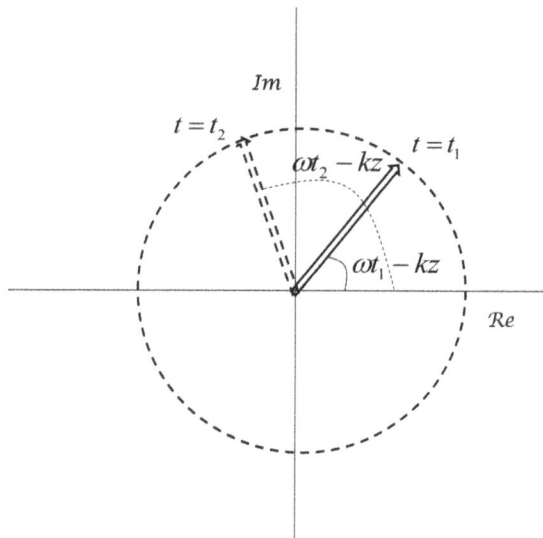

Figure 2.7. Complex notation of waves.

velocity refers to particle (like air molecules) velocity. Similarly, an electromagnetic wave carries the electromagnetic power in the form of electric and magnetic waves. At the boundary of media having different propagation properties, the two quantities independently satisfy the boundary conditions. We will discuss this topic in detail later in section 3.1.2.

Normally, the force-like and velocity-like quantities are connected by a constant. Acoustic pressure and velocity waves are connected by acoustic impedance (a medium constant). Electric and magnetic waves are connected with the speed of light. Thus, for simplicity, we usually interpret that the square of the amplitude is the power density. (Note that dimensional analysis indicates that the product is power density; e.g. pressure times velocity = $(\text{N m}^{-2}) \times (\text{m s}^{-1}) = (\text{N (sm)}^{-1}) = (\text{J (sm}^2)^{-1}) = (\text{W m}^{-2})$. This represents a flow of the power per unit cross-sectional area.) If we use the exponential expression for a wave, we can evaluate the wave power P by multiplying the complex number representing the amplitude $Ae^{i(\omega t - kz)} \equiv Ae^{i\theta}$ to its complex conjugate $A^* = Ae^{-i\theta}$

$$P = AA^* = Ae^{i\theta} Ae^{-i\theta} = A^2. \tag{2.90}$$

This power is twice the average power. The amplitude corresponding to the average power is referred to as the root mean square (RMS). Thus, the RMS value is $1/\sqrt{2}$ of amplitude A (the peak value). We can easily see this by using the sinusoidal form of the wave

$$
\begin{aligned}
P_{av} &= \frac{1}{\tau} \int_0^\tau A^2 \cos^2(\omega t - kz) dt = \frac{1}{2\tau} \int_0^\tau A^2(1 + \cos 2(\omega t - kz)) dt \\
&= \frac{1}{2\tau} A^2 \left[\left(t + \frac{1}{2\omega} \sin 2\left(\omega t - kz\right) \right) \right]_0^\tau = \frac{A^2}{2} = \frac{P}{2}.
\end{aligned}
\tag{2.91}
$$

Therefore, $A_{RMS} = \sqrt{P_{av}} = A/\sqrt{2}$. Here τ is the period; $\tau\omega = 2\pi$.

Phase. Often the phase of a wave is more important than the amplitude. Especially, in the situation where two or more waves are adding to each other, the phase difference between the component waves plays an important role in determining the overall intensity. However, it is not easy to detect a phase directly. To detect the phase difference of a pair of waves, we can look at the intensity of the total wave. From the intensity variation, we can calculate the phase difference. This technique is known as interferometry, and is applied in various fields of science and engineering. Here, we first discuss the phase on the complex plane. In the next section, we will discuss the phase in the context of superposition of waves. Interferometry will be discussed in chapter 3.

Consider the complex notation of a wave function in the form of $\psi_0 \cos(\omega t \pm kz)$ on the complex plane in figure 2.7. Here, the radius represents the amplitude ψ_0 and the angle from the real axis represents the phase. When the amplitude ψ_0 is constant (the amplitude remains the same as the wave travels), figure 2.7 is referred to as the phase diagram.

On a phase diagram, the advancement of the phase is represented by a counter-clockwise movement of the tip of the constant radius. From the physical point of view, the projection to the real axis (the real part) is meaningful as it represents the harmonic signal conveyed by the wave. If the wave function is in the form of $\psi_0 \sin(\omega t \pm kz)$, the imaginary part is physically meaningful. If the wave function has a nonzero initial phase ϕ_0, we can express it as $\psi(t, z) = \psi_0(\cos(\omega t \pm kz) + \phi_0)$. In this case, $\psi(0, 0) = \psi_0 \cos \phi_0$ becomes a complex number with the angle ϕ_0 from the real axis. Although the initial position is not on the real or imaginary axis, the real part represents the cosine variation of the wave function as the phase advances either in time or space.

In section 2.2.1 we discussed that multiplication of the complex unit i to the complex notation of a wave is equivalent to a phase shift of $\pi/2$ (see text near equation (2.39)). We know that differentiation of a sine function results in a cosine function and differentiation of a cosine function results in a negative sine function. We also know that sine and cosine functions have a phase difference of $\pi/2$ and that the temporal differentiation and integration of a harmonic function in the exponential notation are the multiplication and division by $i\omega$, respectively. These all together indicate that somehow we can deal with differentiation of wave functions with the use of complex notation. Indeed, using complex notation, we can conveniently perform differential and integral operations on waves; the differentiation is multiplication of $i\omega$ to the original wave function and the integration is division by the same factor. In other words, for every temporal differentiation, the amplitude is multiplied by a factor of angular frequency and the phase is advanced by $\pi/2$. We can easily understand these by considering the following identities

$$\frac{d}{dt}\cos(\omega t \pm kz) = -\omega\sin(\omega t \pm kz) = \omega\cos\left(\omega t \pm kz + \frac{\pi}{2}\right) \quad (2.92)$$

$$i\omega e^{i(\omega t \pm kz)} = \omega(i\cos(\omega t \pm kz) - \sin(\omega t \pm kz))$$
$$= \omega\left(\cos\left(\omega t \pm kz + \frac{\pi}{2}\right) + i\sin\left(\omega t \pm kz + \frac{\pi}{2}\right)\right). \quad (2.93)$$

The real part of equation (2.93) is identical to equation (2.92), validating the equivalence of temporal differentiation and multiplication of $i\omega$ in the complex notation.

We can interpret the above mathematical formulation from some physical point of view. Say the above $\cos(\omega t - kz)$ is a displacement wave. According to the logic we argued above, the phase of the velocity wave is $\pi/2$ ahead of the displacement wave. Similarly, the phase of the acceleration wave is $\pi/2 \times 2 = \pi$ ahead of the displacement, which means the sign of the acceleration wave is opposite to the displacement wave. As for the amplitude, the velocity wave is $\times\omega$ of the displacement wave and the acceleration wave is $\times\omega^2$ of the displacement wave. We can intuitively understand the fact that higher ω makes the amplitude of the velocity wave greater by knowing that higher frequency causes faster oscillation. In chapter 1, we discussed that the elastic force is proportional to the displacement with a negative sign. Force is proportional to acceleration. Acceleration is the secondary derivative of displacement with respect to time. We can clearly see this pattern in equation (2.93). Multiplying $i\omega$ twice changes the sign in front of the displacement wave.

Repeating the same type of argument, we can understand that temporal integration is equivalent to the division by $i\omega$ as follows

$$\int \cos(\omega t \pm kz)dt = \frac{1}{\omega}\sin(\omega t \pm kz) = \frac{1}{\omega}\cos\left(\omega t \pm kz - \frac{\pi}{2}\right). \quad (2.94)$$

$$\frac{1}{i\omega}e^{i(\omega t \pm kz)} = \frac{1}{\omega}(-i\cos(\omega t \pm kz) + \sin(\omega t \pm kz))$$
$$= \frac{1}{\omega}\left(\cos\left(\omega t \pm kz - \frac{\pi}{2}\right) + i\sin\left(\omega t \pm kz - \frac{\pi}{2}\right)\right). \quad (2.95)$$

As the inverse process of differentiation, integration reduces the amplitude by a factor of ω and delays the phase by $\pi/2$.

The phase plays an important role in the power carried by a wave. Above, we argued that the square of amplitude represents the power as the product of the force-like and velocity-like quantities without paying attention to the phase difference. Here we add the phase to the argument.

Consider that the force wave is behind the displacement wave in phase by ϕ, and calculate the power P'_{av} carried by the two waves

$$P'_{av} = \frac{1}{\tau}\int_0^\tau A^2 \cos(\omega t - kz)\cos(\omega t - kz - \phi)dt$$
$$= \frac{1}{2\tau}\int_0^\tau A^2(\cos(2\omega t - 2kz - \phi) + \cos\phi)dt$$
$$= \frac{1}{2\tau}A^2\left[\left(\frac{1}{2\omega}\sin(2\omega t - 2kz - \phi) + t\cos\phi\right)\right]_0^\tau \quad (2.96)$$
$$= \frac{A^2}{2}\cos\phi = \frac{P}{2}\cos\phi = P_{av}\cos\phi.$$

The factor $\cos\phi$, called the phase factor, is multiplied to the average power P_{av}. Consequently, the average power is reduced. When $\phi = \pi/2$, the average power becomes zero. In a real world situation, this can happen when the ac voltage is delayed by $\pi/2$ from the ac current. If this happens, the receiver cannot receive any power; it all gets reflected. It is necessary to adjust the phase delay. This operation is called the phase matching.

Warning regarding the use of complex notation. As we observed above, complex notation is very useful for differentiation, integration and evaluation of the amplitude and phase of superposed waves. However, this convenience comes from Euler's notation that connects sinusoidal signal of a certain frequency (ω) to complex notation $e^{i\omega t}$. We simply use the facts that differentiation and integration correspond to switching from real to imaginary parts and that the differentiation and integration in the complex format is simple. After making necessary mathematical manipulation, we take the real part as a physically meaningful quantity. It should be noted that in this process we keep dealing with the same frequency. We cannot apply the same usefulness to other frequencies at the same time. For instance, we cannot compute the power from multiplication of the complex notation of a velocity-like quantity and the complex notation of the corresponding force-like quantity. For instance, try to use the complex notations in equation (2.91) to compute the average power

$$A^2 \int_0^\tau e^{2i(\omega t \pm kz)} dt = \frac{A^2}{2i\omega} [e^{2i(\omega t \pm kz)}]_0^\tau = \frac{A^2}{2i\omega} e^{\pm 2ikz} [e^{2i\omega t}]_0^\tau \tag{2.97}$$

$$= \frac{A^2}{2i\omega} e^{\pm 2ikz} [e^{2i\omega \tau} - e^0] = A^2 \frac{1}{2i\omega} e^{\pm 2ikz} \left[e^{2i(2\pi)} - 1 \right] = 0. \tag{2.98}$$

Apparently, the use of complex notation in this fashion always leads to zero average power, which is obviously incorrect. Make sure to use the complex notation appropriately.

2.2.4 Superposition of waves and Fourier transform

When two waves share the same space, the physical quantities of individual waves add to each other. For example, if two sound waves are superposed, the air at each point of the shared space experiences the total displacement and pressure. Two (or multiple) waves can add to each other regardless of the frequency of each individual wave. However, from the phenomenological point of view, there is a significant difference between the cases where the waves being superposed have the same frequency or mutually different frequencies.

Superposition of waves of the same frequency. In this case, the phase difference between the component waves characterizes the superposed wave. When two waves interact with each other, the interaction takes place at the same time and space. So, although the phases of waves vary as a function of both time and space, we can use a common set of space–time coordinates for the phase of the component waves. This also means that as long as the frequency is the same for the component waves, the phase difference remains the same.

Consider that the following two waves ψ_1 and ψ_2 are being superposed

$$\psi_1 = A_1 \cos(\omega t - kz) \equiv A_1 \cos \theta \tag{2.99}$$

$$\psi_2 = A_2 \cos(\omega t - kz + \phi) \equiv A_1 \cos(\theta + \phi). \tag{2.100}$$

For simplicity, we use θ to express the common part of the two phases as $\theta = \omega t - kz$, and ϕ to represent the constant phase difference.

We can conveniently find the magnitude and phase of the superposed wave on a complex plane using Euler's notation (2.38). In figure 2.8, $\psi_1 = A_1 e^{i\theta}$ and

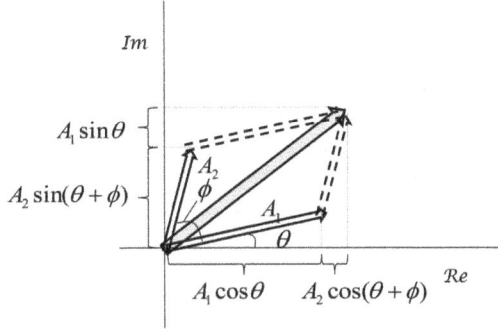

Figure 2.8. Vector additions of waves.

$\psi_2 = A_2 e^{i(\theta+\phi)}$ represent the two waves. According to the complex theory [2], these two complex numbers behave like vectors. Figure 2.8 indicates the vectors resulting from the addition. The resultant vector has one magnitude and phase. Numerically, the amplitude A_{add} and phase θ_{add} of the resultant vector, i.e., the superposed wave, are as follows

$$\begin{aligned} A_{\text{add}}^2 &= (A_1 \cos \theta + A_2 \cos(\theta + \phi))^2 + (A_1 \sin \theta + A_2 \sin(\theta + \phi))^2 \\ &= A_1^2 + A_2^2 + 2A_1A_2(\cos \theta \cos(\theta + \phi) + \sin \theta \sin(\theta + \phi)) \quad (2.101) \\ &= A_1^2 + A_2^2 + 2A_1A_2 \cos \phi \end{aligned}$$

$$\theta_{\text{add}} = \tan^{-1}\left(\frac{A_1 \sin \theta + A_2 \sin(\theta + \phi)}{A_1 \cos \theta + A_2 \cos(\theta + \phi)}\right). \quad (2.102)$$

Note that the magnitude of the superposed wave varies depending on the phase difference ϕ. Figure 2.9 illustrates two sample cases where (a) $\phi = 0.125\pi$ and (b) $\phi = \pi$. In both cases, the amplitude of the first wave is 1 and that of the other wave is 1.4. In the case of (a), the addition of the two waves results in a greater amplitude, whereas in the case of (b) the addition reduces the amplitude. A situation like (a) is referred to as constructive interference of the two waves and a situation like (b) is referred to as destructive interference.

Generally, the magnitude of the superposed wave varies from the maximum value of $A_1^2 + A_2^2 + 2A_1A_2 = (A_1 + A_2)^2$ to the minimum value of $A_1^2 + A_2^2 - 2A_1A_2$ $= (A_1 - A_2)^2$, depending on the phase difference ϕ. The maximum magnitude occurs when the phase difference is such that $\cos \phi = 1$ and the minimum magnitude occurs when $\cos \phi = -1$. Experimentally, we can detect the phase difference ϕ by analyzing the intensity reduction due to the $2A_1A_2 \cos \phi$ term. This is the basics of interferometry, which we will discuss in chapter 3. Figure 2.10 illustrates the variation of the magnitude of the superposed wave as a function of phase difference ϕ for the same component waves as figure 2.9. We can see that the interference is most constructive when $\phi = 0, 2\pi$ and most destructive when $\phi = \pi$. When the amplitude of the component waves are the same, the interference becomes totally destructive for $\phi = \pi$ making the intensity of the superposed wave null.

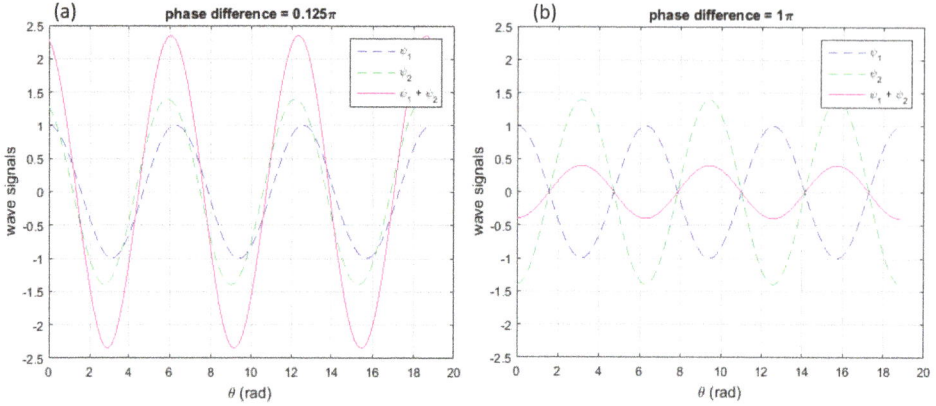

Figure 2.9. Sample wave superpositions with different phase difference ϕ (a): $\phi = 0.125\pi$ and (b): $\phi = \pi$.

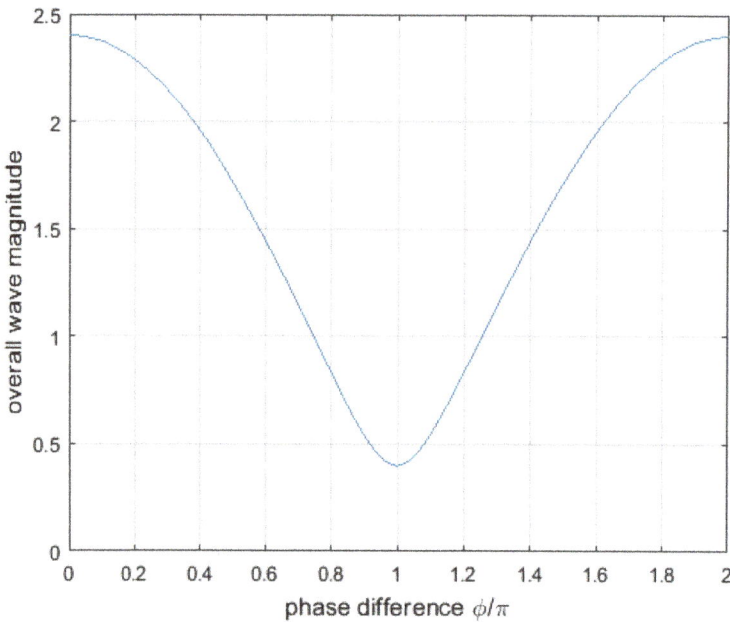

Figure 2.10. Variation of overall magnitude as a function of phase difference. Component waves are the same as in figure 2.9.

Notice that the argument here applies to cases where more than two waves are involved because we can add one wave at a time. Figure 2.11 illustrates the situation schematically. Every time we add a vector representing a wave, we can find the overall amplitude and phase for the summed vector. Therefore, no matter how many waves are superposed, the resultant wave can be characterized by one amplitude and one phase. Diffused reflection of light known as speckles results from superposition of a number of coherent light rays. Figure 2.12 shows a sample image of speckles resulting from diffused reflection of a laser beam on a rough surface. The bright

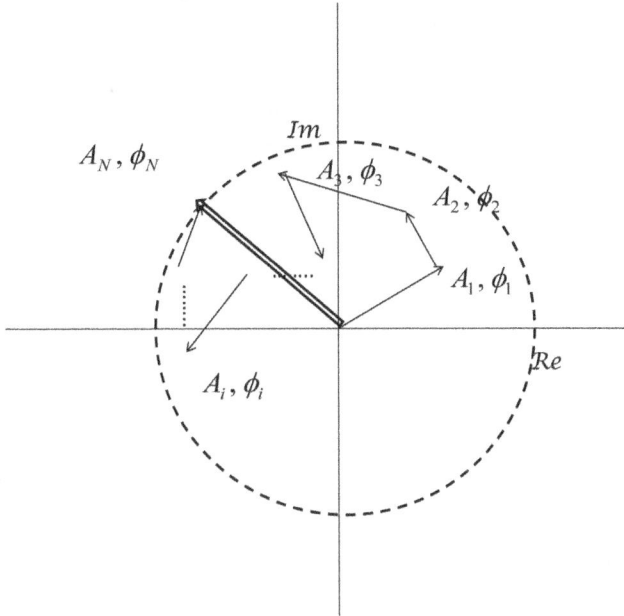

Figure 2.11. Vector additions of many waves; speckles.

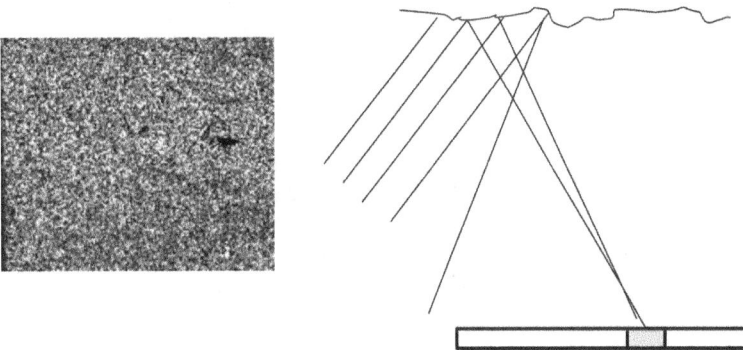

Figure 2.12. Image of speckles (left) and the mechanism of formation (right). A number of rays reflected off a rough surface fall into the same spot of an imaging device. The resultant optical field of each speckle has a definite phase.

spots result from constructive interference of many rays from the surface and the dark spots result from destructive interferences. Although these spots appear to be random, since the speckle pattern results from superposition of coherent sources, a given speckle has one definite phase. This fact is used in an interferometric technique known as electronic speckle-pattern interferometry (ESPI) [6]. ESPI identifies all the points of an object by the phase of speckles.

Superposition of waves of different frequencies. When waves of different frequencies are superposed, unlike the case we observed in the preceding section, the resultant wave cannot be characterized by a single frequency. When two waves of different frequencies are added, we can easily find the temporal behavior of the resultant wave from the trigonometric sum-to-product identity. Consider adding the following two waves

$$\psi_1 = \cos(\omega_1 t - k_1 x) \tag{2.103}$$

$$\psi_2 = \cos(\omega_2 t - k_2 x). \tag{2.104}$$

For simplicity, I am using the same amplitude of unity for the two waves. Using the trigonometric identity, we can express the resultant wave as follows.

$$
\begin{aligned}
\psi_{\text{add}} &= \psi_1 + \psi_2 \\
&= 2 \cos\left(\frac{(\omega_1 + \omega_2)t - (k_1 + k_2)x}{2}\right) \cos\left(\frac{(\omega_1 - \omega_2)t - (k_1 - k_2)x}{2}\right) \quad (2.105) \\
&= 2 \cos(\bar{\omega} t - \bar{k} x) \cos(\omega_m t - k_m x).
\end{aligned}
$$

Here $\bar{\omega}$, $\bar{k}$, ω_m and k_m are defined as follows

$$\bar{\omega} \equiv \frac{1}{2}(\omega_1 + \omega_2), \ \ \bar{k} \equiv \frac{1}{2}(k_1 + k_2) \tag{2.106}$$

$$\omega_m \equiv \frac{1}{2}(\omega_1 - \omega_2), \ \ k_m \equiv \frac{1}{2}(k_1 - k_2). \tag{2.107}$$

Equation (2.106) is the average angular frequency and wave number, whereas equation (2.107) is the differential angular frequency and wave number. Since $\bar{\omega} > \omega_m$ the fist cosine term on the right-hand side of equation (2.105) represents the faster varying feature of the overall wave. The other cosine term represents a slower varying feature, which can be viewed as temporal modulation of the amplitude for the other term. The subscript m denotes modulation. Thus, we can write equation (2.105) in the following form

$$\psi_{\text{add}} = \psi_{m0}(t, x) \cos(\bar{\omega} t - \bar{k} x). \tag{2.108}$$

Here $\psi_{m0}(t, x)$ is the modulated amplitude

$$\psi_{m0}(t, x) \equiv 2 \cos(\omega_m t - k_m x). \tag{2.109}$$

When ω_1 and ω_2 are close to each other, the modulation frequency ω_m is small, and consequently, the amplitude modulation is slow. This low modulation frequency is referred to as the beat frequency, and the corresponding slowly-amplitude-modulated wave phenomenon is referred to as a beat. Figure 2.13 illustrates two sample beats. The top graph is the case where the two frequencies are different by 10% and the bottom one is the case where the frequency difference is 5%. The overall wave form exhibits the fast oscillation representing $\bar{\omega}$ and amplitude modulation ψ_{m0}.

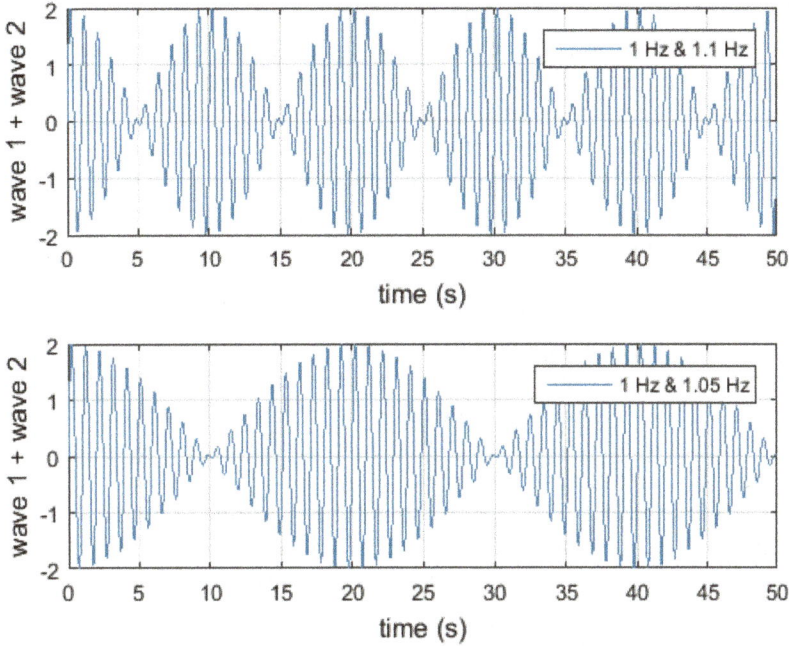

Figure 2.13. Beats.

Sometimes the fast oscillation is called the carrier wave oscillation amd the modulated amplitude is called the (beat) envelope. Careful observation of the top graph will reveal that there are 10 cycles of oscillation in 10 s and five envelopes in 50 s. The former represents the carrier frequency of 1 Hz and the latter the beat frequency of 0.1 Hz. The frequencies of the component waves in this example are 1.1 Hz and 1 Hz.

When a beat phenomenon occurs in a medium where the phase velocity depends on the frequency, the two component waves travel at different velocities. In this case, the envelope travels at a different velocity than the component waves. The phenomenon where the phase velocity depends on the frequency is known as dispersion, and media that cause dispersion are referred to as dispersive media. We will discuss dispersion in detail in the next chapter. The velocity of the envelope in a dispersive medium is called the group velocity. We will discuss group velocity in chapter 4.

Superposition of multiple waves and Fourier transform. When multiple numbers of waves having mutually different frequencies are superposed, the resultant waveform can be completely different from component waves. It is impossible to discuss this subject in general, but I would like to point out its connection with Fourier transform discussed in chapter 1. In short, we can view Fourier transform of a function as superposition of a number of harmonics of sine and cosine functions that constitute the basis of the Fourier expansion. Below, we discuss Fourier transform from the viewpoint of superposition of harmonic functions of different frequencies.

In chapter 1, we discussed that Fourier theorem states that most periodic functions can be expressed with harmonics of cosine and sine functions (frequency components), and I briefly mentioned that Fourier transform is the extension of Fourier series to the limit where the function has an infinite period. The operation of Fourier transform is to find the coefficient for each frequency component. If we use up to Nth harmonics for the frequency components Fourier transform provides us with a set of $2N + 1$ coefficients (one for a constant term plus N for each of the cosine and sine harmonics). In the limit of infinite period, the lowest frequency, and hence the frequency increment[5], becomes infinitesimally small. (We can intuitively understand this by remembering that frequency is reciprocal to the period; if the period is infinite, the corresponding frequency is infinitesimally small.) With the decrease in the frequency increment, on the other hand, the number of harmonics increases. Thus, the set of $2N + 1$ coefficients becomes a continuous function of frequency. These arguments indicate that Fourier transform is closely related to superposition of an infinite number of basis functions.

You may think that Fourier transform is for a function of time or space, not for waves. However, as I mentioned above, when waves are superposed we observe the resultant phenomena at a certain time and space. So, when we consider waves of multiple frequencies being superposed on a plane of observation, we in fact deal with the spatial distribution of the field. Indeed, under certain conditions, the addition of waves represents the exact Fourier transform of a function. Here, as an example, I would like to briefly discuss the phenomenon known as Fraunhofer diffraction of light.

Fraunhofer diffraction is far-field diffraction [7]. Figure 2.14 illustrates a far-field diffraction pattern of light formed by a slit covered with a semi-transparent film. A light wave of a plane wavefront passes through the slit from the left in the positive x direction. The slit is a few centimeters in length along the y-axis and its width along the z-axis is several hundred times the wavelength. According to Huygens's principle [8], we can visualize the situation as a number of spherical wavelets across the slit width emitting rays in all directions. Here, each wavelet has a different intensity, depending on the transmission of the film as a function of z (called the aperture function). We can interpret the direction of each ray as the propagation vector of a plane wave traveling in that direction. On the image plane, each of these propagation vectors has a different projection, which can be interpreted as the spatial frequency. Thus, if the image plane is placed at a far distance away from the slit (so that the diffraction is Fraunhofer diffraction), the optical field at different locations can be viewed as a Fourier component of the aperture function.

By placing positive lenses before and after the slit (figure 2.15), we can form a Fraunhofer diffraction pattern at a convenient distance from the slit using a point light source. Here the first lens makes light waves from the point source plane waves, and the second lens forms the diffraction pattern at the focal plane. On the focal

---

[5] The frequency of Nth harmonic is N times the lowest frequency as $f_N = Nf_0$. Therefore, the frequency increment $f_{N+1} - f_N = (N + 1 - N)f_0 = f_0$ becomes infinitesimally small.

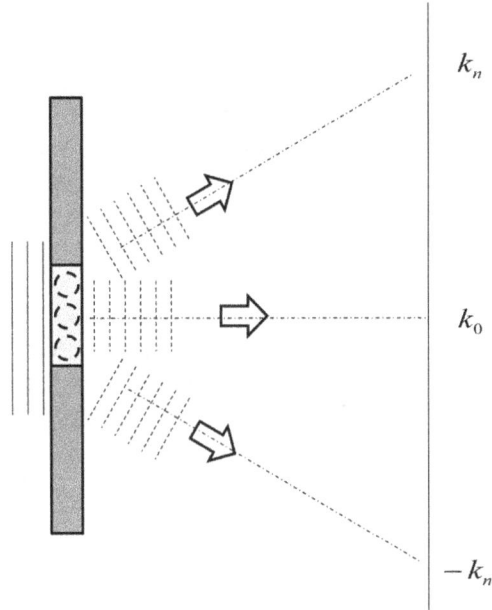

Figure 2.14. Fraunhofer diffraction pattern as Fourier transform.

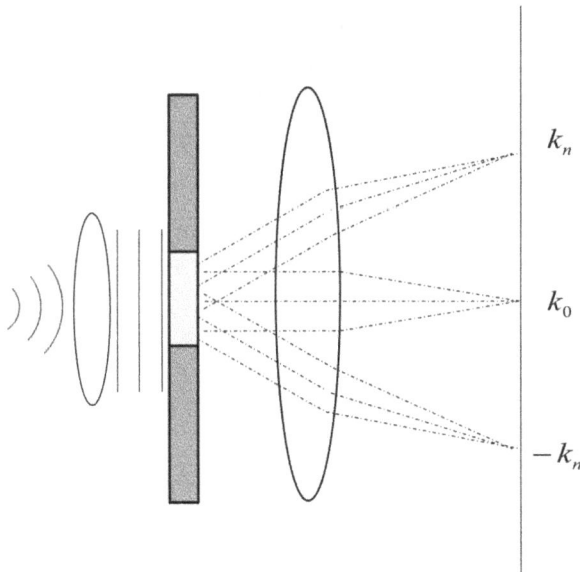

Figure 2.15. Fraunhofer diffraction pattern imaged with a positive lens.

plane, a positive lens converges parallel rays on a single point at a certain distance from the optical axis depending on their angle of propagation to the optical axis.

Numerical demonstration of inverse Fourier transform. Next, consider superposition of harmonics in the context of numerical analysis. The above argument indicates that we can reproduce a given function by summing all the frequency components with the proper coefficients (i.e. inverse Fourier transform). Ideally, we should use infinite numbers of frequency components (harmonics) to reproduce the function. In a real world situation, we cannot add an infinite number of harmonics. We can easily imagine that as the number of harmonics increases the more accurately the inverse Fourier transform represents the function. Here we consider the effect of the number of harmonics on the accuracy of reproduction of a function, using an impulse function as an example.

It is well known that an impulse function in the time (or space) domain yields a flat spectrum in the frequency domain. This indicates that if we add harmonics with the same weight (i.e. take a linear combination of all frequency components with a coefficient of one), we can reproduce an impulse function. In doing so, if we involve the greater number of harmonics the result of the superposition becomes closer to the impulse function. We can easily demonstrate this by actually reproducing a delta function from a number of harmonics of sine and cosine functions. Here, I would like to make a simple demonstration. The demonstration is as follows. I define a limit of x as $-0.25 < x < 0.25$ with $\Delta x = 0.0001$, and reproduce an impulse (a delta) function $\delta(x)$. The fundamental frequency in this series is 1 as it corresponds to a half period of $0.25-(-0.25) = 0.5$, see figure 2.16. Since $\delta(x)$ is an even function in the defined range of x, we can reproduce the impulse function with cosine terms only. Here, as is clear from the above argument, the greater the number of harmonics we use, the better we can reproduce the impulse function.

Figure 2.17 shows two cases of reproduction. The top graph is when we use up to 100th harmonics of the cosine series and the bottom one is when we use up to 1000th harmonics. The peak value of the reproduced function is normalized to one. We can

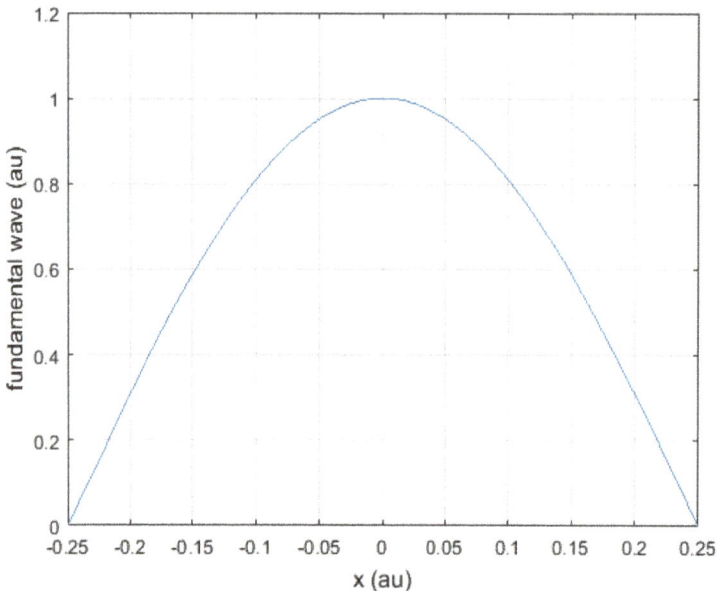

Figure 2.16. Fundamental cosine function to reproduce impulse function.

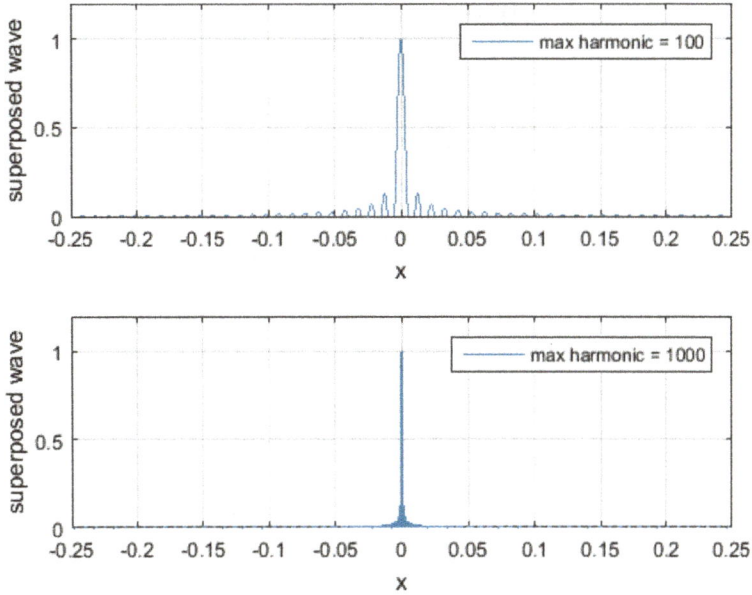

Figure 2.17. Reproduction of impulse function.

clearly see that the 1000th terms case reproduces the impulse function more accurately. The oscillatory behavior observed near $x = 0$ in the 100 terms case is due to the lack of the number of harmonics.

References

[1] Spencer A J M 1980 *Continuum Mechanics* (New York: Longman)
[2] Boas M L 2006 *Mathematical Methods in the Physical Sciences* 3rd edn (London: Wiley) ch 2
[3] Boas M L 2006 *Mathematical Methods in the Physical Sciences* 3rd edn (London: Wiley) ch 13
[4] Yariv A 1971 *Introduction to Optical Electronics* (New York: Holt, Rinehart and Winston)
[5] Schmerr L W Jr 2016 *Fundamentals of Ultrasonic Nondestructive Evaluation: A Modeling Approach* 2nd edn (Basel: Springer)
[6] Meinlschmidt P, Hinsch K D, Sirohi R S (eds) 1996 *Selected Papers on Electronic Speckle Pattern Interferometry Principles and Practice*, Milestone Series, vol 132 (Bellingham, WA: SPIE)
[7] Hecht E 2002 *Optics* 4th edn (San Francisco, CA: Addison Wesley) ch 10 and 11
[8] Hecht E 2002 *Optics* 4th edn (San Francisco, CA: Addison Wesley) ch 4

Chapter 3

Basic properties of waves

3.1 Reflection and refraction

When a wave comes to a boundary of different media, the following two things happen. (a) Part of the wave starts to propagate backward. (b) If the incident wave is not normal to the boundary surface, the transmitted wave changes its direction. (a) is known as reflection and (b) as refraction. Both phenomena occur commonly in every wave. Apparently, two questions are raised. In what direction do the reflected and transmitted wave go, and what percentage of the energy in the incident wave is reflected or transmitted? The first question is answered by the laws of reflection and refraction. The answer to the second question is determined by the boundary conditions of a given system that satisfy certain physical laws. Below we will discuss answers to these questions.

3.1.1 Angles of reflection and refraction

In short, reflection and refraction occur by the following mechanism. As discussed on various occasions in this book (e.g. section 2.1.3), the medium determines the wave velocity depending on its medium constants (e.g. the elastic constant and density). Therefore, as the wave passes through a boundary, it changes the velocity according to the change in the medium constants. The frequency of the wave, on the other hand, is determined by the source. Consequently, when a wave is incident to a boundary, the reflected wave keeps the same wavelength as the incident wave (because they are in the same medium), whereas the transmitted wave has a different wavelength. At the boundary this condition is somehow established. As we will see in the following sections, this is true only when the reflected and transmitted waves propagate in certain directions. The laws governing this situation are referred to as the laws of reflection and refraction.

doi:10.1088/978-1-6817-4573-2ch3
3-1

It is convenient to introduce the quantity known as the index of refraction n defined as the ratio of the wave velocity in air to a medium[1]

$$n_1 = \frac{v_0}{v_1} \tag{3.1}$$

$$n_2 = \frac{v_0}{v_2}. \tag{3.2}$$

Here n_1 and n_2 are the index of refraction in medium 1 and 2, v_1 and v_2 are the wave velocity in medium 1 and 2, and v_0 is the wave velocity in air. In the present context, medium 1 and 2 are on the opposite sides of the boundary. Since the frequency ν is common, we can express the wavelength in each medium using the index of refraction as follows

$$\lambda_1 = \frac{v_1}{\nu} = \frac{v_0/n_1}{\nu} = \frac{v_0/\nu}{n_1} = \frac{\lambda_0}{n_1} \tag{3.3}$$

$$\lambda_2 = \frac{v_2}{\nu} = \frac{v_0/n_2}{\nu} = \frac{v_0/\nu}{n_2} = \frac{\lambda_0}{n_2}. \tag{3.4}$$

Here equations (3.1) and (3.2) are used in going through the second equal sign of equations (3.3) and (3.4), and λ_0 is the wavelength in air.

Consider in figure 3.1 that a wave is approaching the boundary that separates the top medium (index of refraction n_1) and the bottom medium (index of refraction n_2). Here solid lines $S_A A$ and $S_B B$ are two representative rays (propagation vectors) of the incident wave[2] and the dashed lines between them are two representative wavefronts that are exactly one wavelength apart. The angle made by the propagation vectors and the normal vector of the surface (N_A) is θ_1. Assume that these rays reflect off the boundary at an angle θ_2 as represented by dot–dashed lines AR_A and BR_B.

In triangles ABD and BAC we find that lines DB and AC, respectively, are equal to the wavelength before and after the reflection. Since both the incident and reflected waves are in the same medium whose index of refraction is n_1, the wavelength is commonly λ_1. Hence, we can derive the following conditions

$$DB = AB \sin \theta_1 = \lambda_1. \tag{3.5}$$

---

[1] In the case of an electromagnetic wave, usually the index of refraction is defined as the ratio of the wave velocity in a medium to that in vacuum. However, since under normal conditions the wave velocity in air is practically the same as in vacuum, we can define the index of refraction as the ratio of the wave velocity in air to a medium.

[2] We can consider that these two lines define the spot size of a Gaussian beam (the point where the amplitude is $1/e$ of the beam center). However, the use of any two propagation vectors drawn in the incident wave holds the same argument.

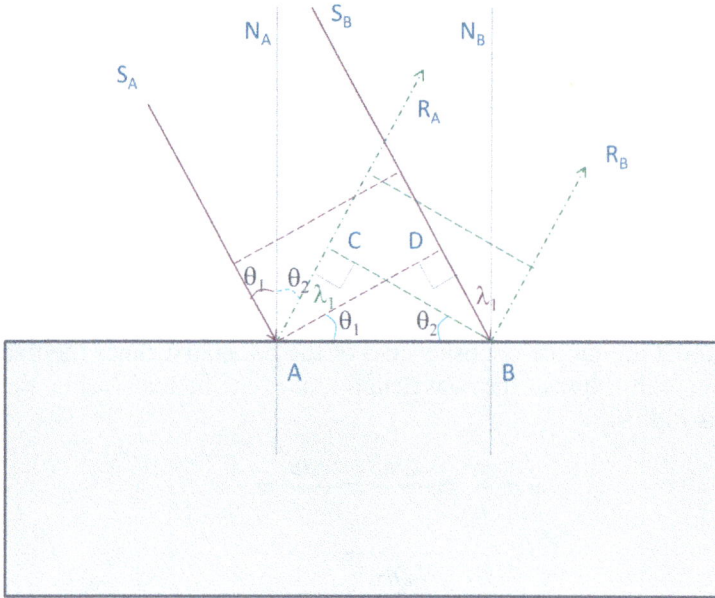

Figure 3.1. A wave reflected off a boundary.

$$AC = AB \sin \theta_2 = \lambda_1. \tag{3.6}$$

From equations (3.5) and (3.6),

$$\sin \theta_1 = \sin \theta_2. \tag{3.7}$$

Since $\theta_1 < 90°$ and $\theta_2 < 90°$, it follows

$$\theta_1 = \theta_2. \tag{3.8}$$

The equality (3.8), referred to as the law of reflection, states that the angle of reflection is the same as the angle of incidence.

Now consider in figure 3.2 the transmitted portion of the wave. Call the angle made by the propagation vector of the transmitted wave and the normal vector (N_A) the angle of refraction, θ_2'. Repeat the same argument made above for reflection, and find the following relation between the angle of incidence and refraction

$$AB \sin \theta_1 = \lambda_1 \tag{3.9}$$

$$AB \sin \theta_2' = \lambda_2. \tag{3.10}$$

Substituting equations (3.3) and (3.4) into equations (3.9) and (3.10), we obtain the following relation between angles θ_1 and θ_2'

$$n_1 \sin \theta_1 = n_2 \sin \theta_2'. \tag{3.11}$$

Equality represented by equation (3.11) is known as the law of refraction or Snell's law [1–3].

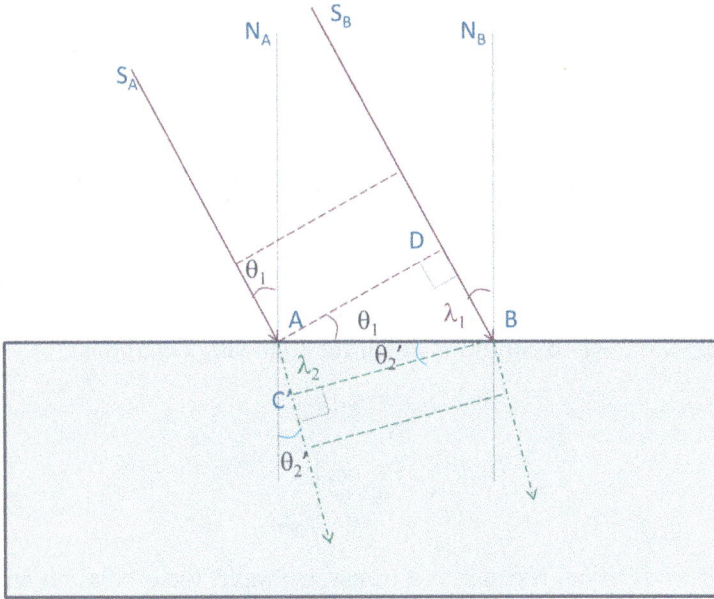

Figure 3.2. A wave transmitted through a boundary.

Equation (3.11) indicates that if a wave is incident to a medium whose index of refraction is greater ($n_1 < n_2$), the angle of refraction is smaller than the angle of incidence, $\theta_2' < \theta_1$. Reflection under this condition is referred to as external reflection. External reflection is always accompanied by transmission; we can always find the angle of refraction with equation (3.11).

If a wave is incident to a medium whose index of refraction is smaller ($n_1 > n_2$), the angle of refraction is greater than the angle of incidence, $\theta_2' > \theta_1$. Reflection under this condition is referred to as internal reflection. In this case, as the angle of incidence increases, it comes to the point where the angle of refraction reaches 90°. If the angle of incidence is further increased, it becomes impossible to find the angle of refraction by equation (3.11). In this situation, there is no transmission. Reflection under this condition is referred to as total reflection, and the angle of incidence that makes the angle of refraction equal to 90° is referred to as the critical angle, θ_c. Any angle of incidence greater than the critical angle makes the reflection total reflection. From equation (3.11), we find

$$\sin \theta_c = \frac{n_2}{n_1}. \tag{3.12}$$

3.1.2 Coefficient of reflection and transmission

In this section, we discuss what proportion of an incident wave is reflected (the coefficient of reflection) or transmitted through (the coefficient of refraction) the boundary. The coefficients of reflection and refraction are determined by the boundary conditions. Since the boundary conditions result from the specific law of physics that governs the physical quantity propagating as the wave, the coefficients of reflection and refraction must be discussed independently for each type of wave. Below we start off the discussion for a wave on a string considering the underlying physics in some detail, and continue to other types of waves.

Boundary conditions for a wave on connected string. First we discuss reflection and transmission of a wave on a string, a transverse wave we discussed in section 1.1. Consider a string connected at its right end to another string of different linear density (m/l in equation (1.80)), and a wave traveling from the left to the right on the connected strings. At the boundary, the following two conditions must hold. (a) the vertical displacement of the string is continuous; (b) the vertical force is continuous. We can express these two conditions with the following equations

$$(a)\ \xi_y|_{\text{left}} = \xi_y|_{\text{right}} \tag{3.13}$$

$$(b)\ \frac{d\xi_y}{dx}\bigg|_{\text{left}} = \frac{d\xi_y}{dx}\bigg|_{\text{right}}. \tag{3.14}$$

Condition (a) is trivial because as a transverse wave, the oscillation amplitude of the wave should be the same on both sides of the boundary. In other words, the left-hand and right-hand sides of equation (3.13) represent the same displacement of the string (sometimes this is referred to as the particle displacement) viewed from the left and right sides of the boundary[3].

Condition (b) is also trivial if we recall equation (1.78), which represents the net force acting on the infinitesimal portion of the string dx between $x = x + dx$ and $x = x$

$$F_y^{\text{net}} = T\left(\frac{d\xi_y}{dx}\bigg|_{x+dx} - \frac{d\xi_y}{dx}\bigg|_x\right). \tag{1.78}$$

At the boundary, $dx = 0$ and hence there is no net force, i.e., $F_y = 0$.

It is instructive to consider conditions (a) and (b) from the physical point of view. (a) describes the boundary condition regarding the displacement and (b) regarding the force. We know that the product of force and displacement represents energy. This intuitively indicates that the combination of the two conditions somehow describes the behavior of the wave motion in association with energy. Indeed, as we will find soon, they represent conservation of energy at the boundary. The fact that the two conditions are satisfied independently of each other indicates it is worthwhile discussing the wave behavior of the displacement and force separately. Below, we will discuss the boundary behavior of the displacement wave first, followed by that of the force wave.

---

[3] Sometimes the continuity of the particle velocity instead of displacement is used. On both sides of the boundary the particle velocity is defined as the displacement per unit time. Therefore, as far as the boundary condition is concerned, the arguments based on the displacement and velocity are equivalent to each other.

Displacement wave on a string. In section 1.4.1, we found that a string wave solution in the form of a sinusoidal function (equation (1.82)). Let's put the displacement wave in the exponential form as follows and continue the discussion. Equation (1.83) indicates that the wave velocity is inversely proportional to the linear density of the string. Since the linear density of the string is different on the opposite sides of the boundary, the wave velocity changes across the boundary. However, the frequency of the oscillation is the same on both sides, as at the boundary the string oscillates at a certain rate. Consequently, the wavelengths or the wave numbers are different on the opposite sides of the boundary. Expressing the oscillation amplitude and wave number of the left and right strings with subscript 1 and 2, respectively, we can express the incident, reflected and transmitted displacement waves as follows.

$$\xi_i = \xi_{i0}e^{i(\omega t - k_1 x)} \tag{3.15}$$

$$\xi_r = \xi_{r0}e^{i(\omega t + k_1 x)} \tag{3.16}$$

$$\xi_t = \xi_{t0}e^{i(\omega t - k_2 x)}. \tag{3.17}$$

Here ξ_{i0} etc are the vertical oscillation amplitudes for the respective displacement waves. The overall displacement wave at the boundary is defined as ξ_y in equation (3.13). Here, we drop the subscript y for simplicity. Note that the incident and transmitted waves propagate in the positive x-direction and the reflected wave in the negative x-direction.

Setting $x = 0$ at the boundary and using equations (3.15)–(3.17), we can express the two conditions (3.13) and (3.14) as follows

$$\xi_{i0} + \xi_{r0} = \xi_{t0} \tag{3.18}$$

$$k_1\xi_{i0} - k_1\xi_{r0} = k_2\xi_{t0}. \tag{3.19}$$

Elimination of ξ_{t0} and ξ_{r0} yields the following expressions for the reflection and transmission coefficients R_ξ and T_ξ

$$R_\xi \equiv \frac{\xi_{r0}}{\xi_{i0}} = \frac{k_1 - k_2}{k_1 + k_2} \tag{3.20}$$

$$T_\xi \equiv \frac{\xi_{t0}}{\xi_{i0}} = \frac{2k_1}{k_1 + k_2}. \tag{3.21}$$

Recalling that the wave (phase) velocity is $v_p = \omega/k$ (equation (2.36)), we can rewrite equations (3.20) and (3.21) in terms of v_p as follows

$$R_\xi = \frac{v_{p2} - v_{p1}}{v_{p2} + v_{p1}} \tag{3.22}$$

$$T_\xi = \frac{2v_{p2}}{v_{p2} + v_{p1}}. \tag{3.23}$$

Equation (3.22) indicates that if $v_{p2} < v_{p1}$ the reflection coefficient R_ξ is negative. According to equation (2.36), v_p is inversely proportional to the square root of the linear density; the heavier the string the slower the wave travels. We can interpret the negative R_ξ as follows. When an incident wave comes to a boundary entering a heavier string, the oscillation amplitude on the front (the heavier string) side is smaller because the force (which is the same on both sides of the boundary according to condition (b)) has to work on the greater mass; Newton's second law says that the resultant acceleration, hence the displacement in the oscillatory dynamics, becomes less. Condition (a), on the other hand, requires that the overall oscillation amplitude must be the same on both sides. To satisfy both requirements, it is necessary that the reflected wave adds to the incident wave negatively so that the overall amplitude on the incident side be the same as the transmitted side (figure 3.3).

An extreme case is when the second string has infinite mass, or the boundary is fixed. This case is known as the fixed end reflection where the reflected wave completely negates the incident wave so that the total oscillation amplitude at the boundary is null. In terms of the phase, we can say that the reflected wave undergoes a phase shift of π.

The other extreme case is when the second string is massless. This case is known as the free end reflection. There is no phase delay at the free end, and the reflected wave simply changes the direction keeping the pattern of oscillation the same.

Specific impedance. At this point, I would like to introduce the concept known as the specific impedance. As you will easily guess, this concept comes from the impedance z defined in electric circuit theory as $V = zI$, where V is the voltage and I is the current. We can interpret this as 'how much voltage is necessary to supply a certain amount of current?'. In the context of an electric circuit, it represents how many volts you need to provide to your flashlight so that the current flowing through the circuit is sufficient to make the light bright enough. If the dry batteries you put in previously do not provide high enough voltage, you will need to replace them with new ones because you do not have control over the impedance; it is a quantity determined by the system. Similarly, the impedance of your artery decides how much pressure your heart must provide to maintain sufficient blood flow to your brain.

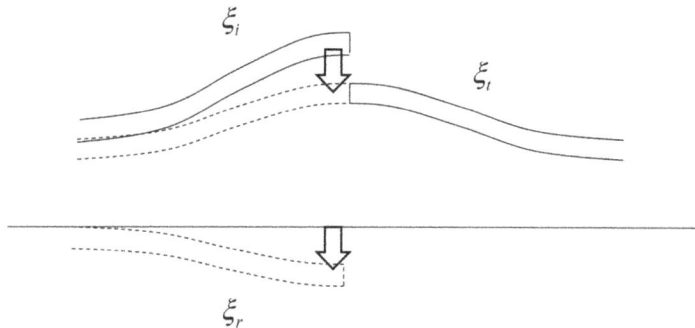

Figure 3.3. A wave on a string at a boundary. ξ_r reduces ξ_i so that at the boundary ξ wave is continuous.

In the above examples, the voltage and pressure represent force on the particle. The pressure generated by the pumping action of the heart exerts force on blood cells (particles) to cause the flow. Similarly, voltage (electrical potential) exerts electric force F_e on charges q via the electric field E (the slope of the electric potential) as $F_e = qE$. Therefore, we can interpret voltage or pressure as a force-like quantity. From these viewpoints, we can understand the impedance as 'force-like quantity = impedance × particle flow'.

In the context of the wave on a string, the impedance is defined as the ratio of the string's elastic force to the velocity. The former corresponds to the force-like quantity and the latter to the particle flow. Using the same exponential form as equations (3.16)–(3.18), we can express the string force (F_y) and the corresponding velocity (both are vertical to the direction of the string wave) using the tension (T) and the small angle approximation ($\sin\theta \cong \tan\theta$) as follows (see equation (1.71))

$$F_y = -T\tan\theta = -T\frac{\partial \xi_y}{\partial x} = ikT\xi_{y0} \tag{3.24}$$

$$v_y = \frac{\partial \xi_y}{\partial t} = i\omega\xi_{y0}. \tag{3.25}$$

Thus

$$z_0 = \frac{F_y}{v_y} = T\frac{k}{\omega} = \frac{T}{v_p}. \tag{3.26}$$

Here v_p is the wave (phase) velocity and $v_p = \omega/k$ is used in the rightmost equality in equation (3.26). As we discussed multiple times, the medium (system) determines the wave velocity. In the present case, $v_p = \sqrt{T/(m/l)}$ (equation (2.36)) consists of medium constants.

Substituting this expression of the wave velocity into equation (3.26), we find the following form of z_0

$$z_0 = \sqrt{T\frac{m}{l}}. \tag{3.27}$$

This form of the impedance expression literally tells us the greater the mass (the inertia), the greater impediment the string becomes.

Now if the medium constant, say the linear density m/l changes, the impedance changes as well. In the context of reflection and transmission of a wave on a string, since the frequency normally remains unchanged, we can express the impedance of medium 1 and 2, z_1 and z_2, as follows. Here we use equation (3.26) with the phase velocity in medium 1 and 2, v_{p1} and v_{p2}

$$z_1 = \frac{T}{v_{p1}} = T\frac{k_1}{\omega} \tag{3.28}$$

$$z_2 = \frac{T}{v_{p2}} = T\frac{k_2}{\omega}. \tag{3.29}$$

Reflection and transmission coefficients of displacement wave on a sting. Substitution of equations (3.28) and (3.29) into equations (3.22) and (3.23) yields the following expressions for the reflection and transmission coefficients

$$R_\xi = \frac{\frac{T}{z_2} - \frac{T}{z_1}}{\frac{T}{z_2} + \frac{T}{z_1}} = \frac{z_1 - z_2}{z_1 + z_2} \tag{3.30}$$

$$T_\xi = \frac{2\frac{T}{z_2}}{\frac{T}{z_2} + \frac{T}{z_1}} = \frac{2z_1}{z_1 + z_2}. \tag{3.31}$$

Equation (3.30) tells us that the greater the difference in the impedance, the greater the reflection. If we want the wave to pass through a boundary efficiently, we need to make the impedances on both sides of the boundary as close as possible to each other. The activity to make the two impedances as close as possible is referred to as impedance matching. This phrase is often heard in acoustic microscopy and high frequency electric circuit theory.

Also notice that if we add 1 to R_ξ the result is equal to T_ξ

$$1 + R_\xi = 1 + \frac{z_1 - z_2}{z_1 + z_2} = \frac{(z_1 + z_2) + (z_1 - z_2)}{z_1 + z_2} = \frac{2z_1}{z_1 + z_2} = T_\xi. \tag{3.32}$$

This can be easily explained as follows. On the first string side of the boundary, the total displacement is $\xi_{i0} + \xi_{r0}$, and on the second string side it is ξ_{t0}. Using R_ξ and T_ξ, we can express the total displacement on the respective sides as follows

$$\xi_{i0} + \xi_{r0} = \xi_{i0}(1 + R_\xi) \tag{3.33}$$

$$\xi_{t0} = \xi_{i0}T_\xi. \tag{3.34}$$

From the boundary condition (3.13), $\xi_{i0} + \xi_{r0} = \xi_{t0}$. This leads to equation (3.32).

Reflection and transmission coefficients of force wave on a sting. So far, we have discussed the reflection and transmission of the displacement wave. We now consider the force wave. Equation (3.14) tells us that the force at the boundary is given by the spatial derivative of the vertical displacement. From equations (3.15)–(3.17) and (1.78), we can find the incidence, reflected and transmitted force waves as follows

$$f_i = T\frac{d\xi_i}{dx} = T(-ik_1)\xi_{i0}e^{i(\omega t - k_1 x)} \equiv f_{i0}e^{i(\omega t - k_1 x)} \tag{3.35}$$

$$f_r = T\frac{d\xi_r}{dx} = T(ik_1)\xi_{r0}e^{i(\omega t + k_1 x)} \equiv f_{r0}e^{i(\omega t + k_1 x)} \tag{3.36}$$

$$f_t = T\frac{d\xi_t}{dx} = T(-ik_2)\xi_{t0}e^{i(\omega t - k_2 x)} \equiv f_{t0}e^{i(\omega t - k_2 x)} \tag{3.37}$$

which leads to

$$f_{i0} = T(-ik_1)\xi_{i0} \tag{3.38}$$

$$f_{r0} = T(ik_1)\xi_{r0} \tag{3.39}$$

$$f_{t0} = T(-ik_2)\xi_{t0}. \tag{3.40}$$

Substituting equations (3.38)–(3.40) into equations (3.18) and (3.19), we find as follows

$$k_2 f_{i0} - k_2 f_{r0} = k_1 f_{t0} \tag{3.41}$$

$$f_{i0} + f_{r0} = f_{t0}. \tag{3.42}$$

Repeating the same argument as the displacement wave, we find the reflection and transmission coefficients for the force wave as follows

$$R_f \equiv \frac{f_{r0}}{f_{i0}} = \frac{k_2 - k_1}{k_1 + k_2} = \frac{z_2 - z_1}{z_1 + z_2} \tag{3.43}$$

$$T_f \equiv \frac{f_{t0}}{f_{i0}} = \frac{2k_2}{k_1 + k_2} = \frac{2z_2}{z_1 + z_2}. \tag{3.44}$$

Also, we can easily prove that

$$1 + R_f = 1 + \frac{z_2 - z_1}{z_1 + z_2} = \frac{(z_2 + z_1) + (z_2 - z_1)}{z_1 + z_2} = \frac{2z_2}{z_1 + z_2} = T_f. \tag{3.45}$$

Equation (3.45) literally indicates that the force at the boundary is continuous, or the boundary condition (3.14) holds.

It is interesting and intuitive to compare the reflection and transmission coefficients of the displacement and force waves. Equations (3.30) and (3.43) indicate that when the second medium is higher in impedance, (a) the reflection coefficient of the displacement wave is negative and (b) that of the force wave is positive. We already reasoned for (a) as follows; when the second medium has a higher impedance, i.e., the higher linear density, the amplitude of the displacement wave reduces as it passes through the boundary. This situation requires that the reflected displacement wave must have an opposite sign to the incident wave so that the boundary condition and Newton's law are satisfied (see the paragraph below equation (3.23)). For the same reason, the amplitude of the force wave must be enhanced at the boundary so that on the transmission side the force wave can do heavy-lifting for the medium of the higher linear density; consequently the sign of the reflected force wave is the same as the incident wave or the reflection coefficient is positive.

There is some more argument. The wave on a string is driven by the elasticity of the string material. As we all well know, the total elastic energy is proportional to the product of the elastic force and the maximum displacement ($(1/2)kA^2$ for a spring oscillation with spring constant k and oscillation amplitude A). This indicates that the product of the displacement and force wave amplitude is proportional to the energy carried by the string. At the boundary, the energy must be conserved. Therefore, as much as the displacement is reduced the force must be enhanced. The situation is similar to that of voltage step-up or down. When you travel in a region where the voltage from the wall plug is higher than your home country, you need to increase the voltage with a step-up transformer. On the secondary side of the transformer, the voltage is higher but in proportion the current is lower.

Further, it follows that the sum of the product of the displacement and force wave reflection coefficients and that of the transmitted coefficients should be equal to unity (meaning that the energy of the incident wave is equal to the energies of the reflected and transmitted waves). Indeed, if we add the absolute value of the product of the displacement and force wave reflection coefficients and that of the transmission coefficients, the answer is equal to unity.

$$|R_\xi R_f| + |T_\xi T_f| = \frac{(z_1 - z_2)^2}{(z_1 + z_2)^2} + \frac{4z_1 z_2}{(z_1 + z_2)^2} = \frac{(z_1 + z_2)^2}{(z_1 + z_2)^2} = 1. \tag{3.46}$$

Sound wave. A sound wave is a compression wave in a medium [4, 5]. When the medium like air is compressed, its elasticity pushes back the external force that causes the compression. By inertia, the compressed volume expands in the next phase, followed by another compressive phase. In this fashion, the cycle of compression and expansion continues. Since this dynamic occurs with a phase delay between the front and back ends of the volume, it propagates as a wave. Naturally, the compression and expansion accompany pressure change. Thus, a sound wave is a wave of compression and at the same time a wave of pressure. The amount of compression for a given pressure depends on the medium's volume expansion coefficient K. The elastic energy determined by the compression and K propagates through the medium.

We first quickly formulate the above dynamics starting from the equation of motion. In figure 3.4 imagine that a block of air is compressed by pressure from the surrounding air. Consequently, its volume changes from the natural volume, the volume with zero pressure, to a smaller volume. We can express the change in the volume using divergence of displacement $\nabla \cdot \boldsymbol{\xi}$. Thus the relationship between the pressure and the volume change becomes

$$p = -K\nabla \cdot \boldsymbol{\xi}. \tag{3.47}$$

Here the negative sign is necessary because pressure causes compression and the divergence operator is defined to be positive for expansion. Equation (3.47) corresponds to Hooke's law applied to a spring, $f = -kx$ where k is the spring constant and x is the stretch of the spring from its natural length.

$$\nabla \cdot \boldsymbol{\xi} = \frac{\partial \xi_x}{\partial x}\hat{x} + \frac{\partial \xi_y}{\partial y}\hat{y} + \frac{\partial \xi_z}{\partial z}\hat{z}$$

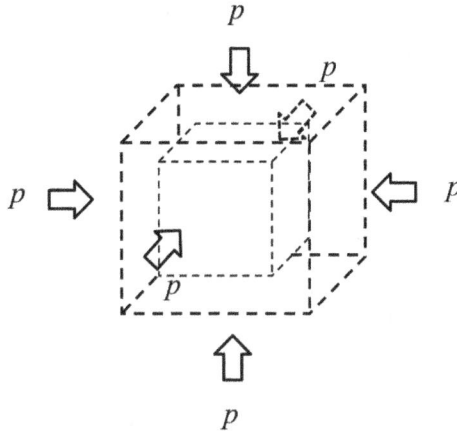

Figure 3.4. A compressed block of air changes its volume.

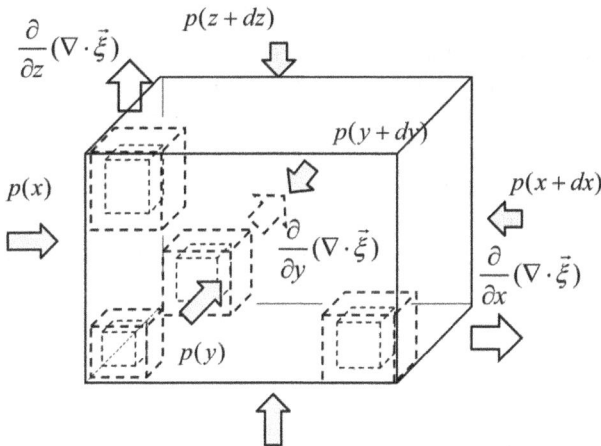

Figure 3.5. Differential pressure accelerates particles in the unit volume.

Now in figure 3.5 consider differential pressure is applied to a unit volume. Being a unit volume, the mass of this volume is equal to the density ρ. Due to the differential pressure, all the particles in the unit volume are accelerated. Thus, we can express the equation of motion as follows

$$\rho \frac{\partial^2 \boldsymbol{\xi}}{\partial t^2} = -\nabla p. \tag{3.48}$$

A negative sign is necessary on the right-hand side of equation (3.48) because a positive pressure gradient causes negative acceleration of displacement (because it would slow down the particle coming out of the volume).

Taking divergence of both sides, we can rewrite equation (3.48) as follows

$$\rho\frac{\partial^2(\nabla \cdot \boldsymbol{\xi})}{\partial t^2} = -\nabla \cdot \nabla p = -\nabla^2 p. \tag{3.49}$$

Substituting equation (3.47) into the left-hand side of equation (3.49), we can eliminate $\boldsymbol{\xi}$ from the equation

$$\frac{\partial^2 p}{\partial t^2} = \frac{K}{\rho}\nabla^2 p. \tag{3.50}$$

Equation (3.50) indicates that pressure p travels as a wave with phase velocity $\sqrt{K/\rho}$.

By substituting equation (3.47) into equation (3.50), we can replace p with $(\nabla \cdot \boldsymbol{\xi})$

$$\frac{\partial^2(\nabla \cdot \boldsymbol{\xi})}{\partial t^2} = \frac{K}{\rho}\nabla^2(\nabla \cdot \boldsymbol{\xi}). \tag{3.51}$$

Equation (3.51) indicates that the volume expansion $(\nabla \cdot \boldsymbol{\xi})$ travels as a wave with phase velocity $\sqrt{K/\rho}$ as well. This wave is referred to as the compression wave of sound.

Boundary conditions for a sound wave. We can argue the reflection and transmission of a sound wave at a boundary in a similar fashion to the case of a wave on a string. At the boundary, a sound wave travels into a medium of different elasticity, density or both. The quantities that constitute the boundary conditions are the particle velocity and the pressure. Between these two quantities, the energy is conserved across the boundary. At the boundary, (a) the velocity component normal to the boundary plane and (b) the pressure must be continuous

$$(a) \quad v_x|_{\text{left}} = v_x|_{\text{right}} \tag{3.52}$$

$$(b) \quad p|_{\text{left}} = p|_{\text{right}}. \tag{3.53}$$

Here v and p are the particle velocity and pressure, and the boundary is a plane normal to the x-axis (figure 3.6).

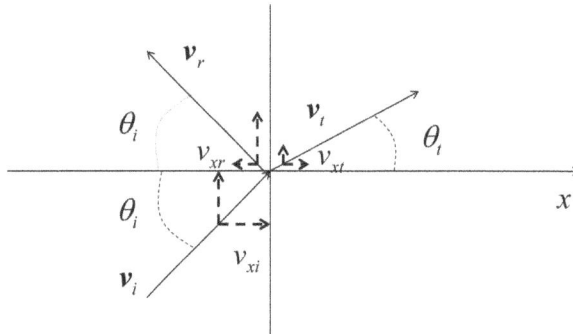

Figure 3.6. A sound wave at a boundary with an oblique incidence angle.

We can understand the two boundary conditions (3.52) and (3.53) as follows. Obviously (a) corresponds to the displacement boundary condition and (b) to the force boundary condition in the case of the string wave discussed above. Condition (b) is trivial. If the pressure is different on the opposite side of the boundary, there must be an external agent applying force on the boundary; this is not the case. Condition (a) may not be obvious but can be explained as follows. Unlike a string wave, a sound wave is a longitudinal wave and can be two- or three-dimensional. If the incident sound wave is normal to the boundary, the particle velocity has only the normal component. This component should be the same whether it is seen from the incident side or transmitted side of the boundary. So, in this case, the particle velocity is continuous. When the incident sound wave is oblique to the plane of boundary, the boundary still oscillates normal to its plane. Thus, the normal component of the particle velocity is continuous.

Acoustic pressure and velocity waves. Expressing the angles of incidence, reflection and refraction with θ_i, $\theta_r(=\theta_i)$ and θ_t, the amplitude of the incident, reflected and transmitted velocity wave with V_i, V_r, and V_t, and the amplitude of the pressure waves with P_i, P_r and P_t, we can rewrite equations (3.52) and (3.53) as follows

$$V_i \cos \theta_i + V_r \cos \theta_i = V_t \cos \theta_t \qquad (3.54)$$

$$P_i + P_r = P_t. \qquad (3.55)$$

By relating the velocity and pressure waves and thereby solving equations (3.54) and (3.55) for V_r and V_t, we can find the expressions of reflection and transmission coefficients for the velocity wave. To this end, let's express both particle velocity $v(t, x)$ and pressure $p(t, x)$ in terms of displacement $\xi(t, x)$. Here we use the same sinusoidal form for the displacement $\xi(t, x) = \xi_0 e^{i(\omega t \pm kx)}$ as equations (3.15)–(3.17) for the wave on a string.

The particle velocity is temporal derivative of the displacement. Therefore,

$$v(t, x) = \frac{\partial \xi}{\partial t} = (i\omega)\xi_0 e^{i(\omega t \pm kx)}. \qquad (3.56)$$

The pressure is due to the force over the unit cross-sectional area. Here the force is elastic force f_{elas} exerted by the medium in proportion to the displacement of the particle

$$f_{\text{elas}} = k \, d\xi = k\frac{\partial \xi}{\partial x}dx = \kappa A\frac{\partial \xi}{\partial x}. \qquad (3.57)$$

Here κ is the elastic constant equivalent to the Young's modulus used in equation (2.11) for the general discussion of longitudinal waves in solids. (Equation (3.57) is the sound wave version of equation (2.10) that relates the elastic force, stiffness and the elastic constant.) Thus, we can express pressure $p = f/A$ as follows

$$p(t, x) = \kappa\frac{\partial \xi}{\partial x} = (\pm ik)\xi_0 e^{i(\omega t \pm kx)}. \qquad (3.58)$$

From Equations (3.56) and (3.58), we can express the incident, reflected and transmitted velocity and pressure waves as follows. These expressions correspond to equations (3.15)–(3.17) and (3.35)–(3.37) for the wave on a string we discussed above for the wave on a string

$$v_i(t, x) = (i\omega)\xi_{i0}e^{i(\omega t - k_1 x)} \equiv V_i e^{i(\omega t - k_1 x)} \tag{3.59}$$

$$p_i(t, x) = (-ik_1)\kappa_1\xi_{i0}e^{i(\omega t - k_1 x)} \equiv P_i e^{i(\omega t - k_1 x)} \tag{3.60}$$

$$v_r(t, x) = (i\omega)\xi_{r0}e^{i(\omega t + k_1 x)} \equiv V_r e^{i(\omega t + k_1 x)} \tag{3.61}$$

$$p_r(t, x) = (ik_1)\kappa_1\xi_{r0}e^{i(\omega t + k_1 x)} \equiv P_r e^{i(\omega t + k_1 x)} \tag{3.62}$$

$$v_t(t, x) = (i\omega)\xi_{t0}e^{i(\omega t - k_2 x)} \equiv V_t e^{i(\omega t - k_2 x)} \tag{3.63}$$

$$p_t(t, x) = (-ik_2)\kappa_2\xi_{t0}e^{i(\omega t - k_2 x)} \equiv P_t e^{i(\omega t - k_2 x)}. \tag{3.64}$$

Here we use subscript 1 and 2 for the medium on the incident and transmitted wave sides, respectively. As in the case of a wave on a string, the phase velocity is different on the incident and transmission sides as v_{p1} and v_{p2}, but the frequency is the same on both sides. Consequently, the wave numbers are $k_1 = \omega/v_{p1}$ and $k_2 = \omega/v_{p2}$.

Acoustic impedance. Similarly to equation (3.26), we can define the impedance for a sound wave. It is obvious that the pressure is the force-like quantity corresponding to F_y in equation (3.26), and the particle velocity is the velocity-like quantity corresponding to v_y in equation (3.26). Further, the phase velocity of a longitudinal wave is given by $v_p = \sqrt{\kappa/\rho}$ (equation (2.41)). We obtain the following expression for the acoustic impedance [6]

$$z = \left|\frac{P}{V}\right| = \kappa\frac{k}{\omega} = \frac{\kappa}{v_p} = \frac{v_p^2\rho}{v_p} = v_p\rho. \tag{3.65}$$

Using equations (3.59)–(3.64) and applying the general expression (3.65) to the medium on the incident wave (subscript 1) and transmitted wave (subscript 2) sides, respectively, we can relate the acoustic pressure and velocity for the incident, reflected and transmitted waves as follows

$$\frac{P_i}{V_i} = \frac{\kappa_1(-ik_1)\xi_{i0}}{i\omega\xi_{i0}} = \frac{\kappa_1(-k_1)}{\omega} = -z_1 \tag{3.66}$$

$$\frac{P_r}{V_r} = \frac{\kappa_1(ik_1)\xi_{r0}}{i\omega\xi_{r0}} = \frac{\kappa_1(k_1)}{\omega} = z_1 \tag{3.67}$$

$$\frac{P_t}{V_t} = \frac{\kappa_1(-ik_2)\xi_{t0}}{i\omega\xi_{t0}} = \frac{\kappa_1(-k_2)}{\omega} = -z_2. \tag{3.68}$$

Note that the incident and transmitted velocity waves have a phase shift of π to the respective pressure waves. The negative sign in front of the impedance in equations (3.66) and (3.68) is because of this fact.

Reflection and transmission coefficients of sound wave. Using equations (3.66)–(3.68), we can rewrite equation (3.55) in the following form

$$z_1(-V_i + V_r) = -z_2 V_t. \tag{3.69}$$

From equations (3.54) and (3.69), we obtain the following expressions of the reflection and transmission for the particle velocity

$$R_v = \frac{z_1 \cos \theta_t - z_2 \cos \theta_i}{z_1 \cos \theta_t + z_2 \cos \theta_i} \tag{3.70}$$

$$T_v = \frac{2z_1 \cos \theta_i}{z_1 \cos \theta_t + z_2 \cos \theta_i}. \tag{3.71}$$

Similarly, by rewriting equation (3.54) with the use of equations (3.66)–(3.68) as follows,

$$\frac{P_i}{z_1} \cos \theta_i - \frac{P_r}{z_1} \cos \theta_i = \frac{P_t}{z_2} \cos \theta_t \tag{3.72}$$

and from equations (3.55) and (3.72), we obtain the reflection and transmission coefficients for the pressure wave

$$R_p = \frac{z_2 \cos \theta_i - z_1 \cos \theta_t}{z_1 \cos \theta_t + z_2 \cos \theta_i} \tag{3.73}$$

$$T_p = \frac{2z_2 \cos \theta_i}{z_1 \cos \theta_t + z_2 \cos \theta_i}. \tag{3.74}$$

We can easily see that the energy conservation (3.46) holds in this case as well

$$|R_v R_p| + |T_v T_p| = \frac{(z_1 \cos \theta_t - z_2 \cos \theta_i)^2}{(z_1 \cos \theta_t + z_2 \cos \theta_i)^2} + \frac{4z_1 z_2}{(z_1 \cos \theta_t + z_2 \cos \theta_i)^2} = 1. \tag{3.75}$$

Electromagnetic wave. An electromagnetic wave is a transverse wave. The electric field vector E and magnetic field vector B are orthogonal to each other, and the wave travels in the direction of $S = E \times H$. Here, S is referred to as Poynting vector, and it carries electromagnetic energy as the wave propagates. H is the magnetic field's auxiliary field vector. In a linear medium, $B = \mu H$ where μ is the magnetic permeability. Below for simplicity and practicality, we continue our discussions for linear media.

Maxwell's equations (3.76)–(3.79) describe the electromagnetic field completely

$$\nabla \cdot \boldsymbol{E} = \frac{\rho}{\varepsilon} \tag{3.76}$$

$$\nabla \times \boldsymbol{E} = -\frac{\partial \boldsymbol{B}}{\partial t} \tag{3.77}$$

$$\nabla \times \boldsymbol{B} = \varepsilon\mu\frac{\partial \boldsymbol{E}}{\partial t} + \mu\boldsymbol{j} \tag{3.78}$$

$$\nabla \cdot \boldsymbol{B} = 0. \tag{3.79}$$

Here $\boldsymbol{E}$ is the electric field vector, $\boldsymbol{B}$ is the magnetic field vector, ρ is the electric charge density, ε is the electric permittivity, and μ is the magnetic permeability. Equation (3.76) indicates how an electric field is generated with electric charges. It is known as Gauss's law. Equation (3.77) and (3.78) are known as Faraday's law and Ampère's law, respectively. They represent the synergetic interaction between the electric and magnetic fields where a temporal change of one induces the other in the form of its rotation. See [7] for more details about Maxwell equations.

Orthogonality of $\boldsymbol{E}$ and $\boldsymbol{B}$ and generation of wave dynamics. Before starting discussions about reflection and transmission of electromagnetic waves, let's take a quick look at Maxwell's equations and see why the electric and magnetic fields are orthogonal to each other, and how they generate wave dynamics. In figure 3.7, consider that the magnetic field $\boldsymbol{B}$ increases downward with time. The cause for the increase can be something such as when the north pole of a bar magnet gets closer to the reference point from above. Equation (3.77) tells us that the temporal change in the magnetic field induces an electric field in the space around, and that the $\partial \boldsymbol{B}/\partial t$ vector is antiparallel to $\nabla \times \boldsymbol{E}$ vector. Figure 3.7 illustrates the situation by expressing the increasing magnetic field with the downward arrow and the induced electric field with the arrows pointing in the tangential directions at several points of a circle around the magnetic field.

By definition, $\nabla \times \boldsymbol{E}$ vector is perpendicular to $\boldsymbol{E}$ vector. In other words, $\boldsymbol{E}$ vectors are in the plane perpendicular to $\nabla \times \boldsymbol{E}$. Temporal differentiation of a vector does not change the spatial orientation of the vector. $\partial B/\partial t$ is antiparallel to $\nabla \times \boldsymbol{E}$. All these pieces of information tell us $\boldsymbol{B} \perp \boldsymbol{E}$.

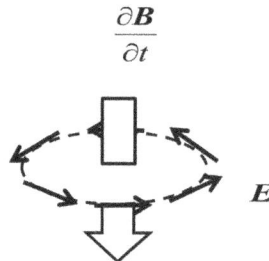

Figure 3.7. Temporal change of magnetic field induces electric field.

Equation (3.78) represents the inverse process in which a temporal change of the electric field induces the magnetic field. Repeating the same thought process as we used for equation (3.77) above, we can easily find that the induced magnetic field is always perpendicular to the electric field.

We can view this induction mechanism as a conversion from temporal differentiation ($\partial/\partial t$) to spatial differentiation ($\nabla \times$). This view naively indicates that this phenomenon represents wave dynamics. Indeed, we can derive wave dynamics from this induction mechanism as follows.

Figure 3.8 illustrates the induction processes for two consecutive time steps (denoted by the superscript). At time step 1, the magnetic field increases downward with time at the origin ($(x, y, z) = (0, 0, 0)$). This temporal change in the magnetic field ($\dot{B}_z^{(1)}$) induces an electric field in the positive y-direction at a point on the x-axis right of the origin, and in the negative y-direction on the x-axis left of the origin. Since this induction takes place at the same time as $\dot{B}_z^{(1)}$, we identify the induced field with superscript 1 as $E_y^{(1)}$. If the rate of the temporal change in the magnetic field is constant, the induced electric field remains as it is.

The wave dynamics occurs when the rate of the temporal change in the magnetic field varies with time. Let's continue the analysis under the condition that the rate of temporal change in B_z keeps increasing downward with time. Because $\ddot{B}_z^{(2)} \neq 0$, $\dot{E}_y^{(2)} \neq 0$. Here we use superscript 2 for $\dot{E}_y$ because the temporal change involves some time from time step 1. Since $\dot{B}_z$ keeps increasing in the same direction, the direction of $\dot{E}_y^{(2)}$ is the same as $E_y^{(1)}$. Thus, $\dot{E}_y^{(2)}$ is into the page on the right of the origin and out of the page on the left. According to equation (3.78), the $\dot{E}_y^{(2)}$ induces a magnetic field $B^{(2)}$ in the zx plane. Here, the into-the-page $\dot{E}_y^{(2)}$ induces a magnetic field clockwise and the out-of-the-page $\dot{E}_y^{(2)}$ induces a magnetic field counterclockwise (figure 3.8).

Very importantly, at the origin, the z-component of $B^{(2)}$ belonging to the clockwise and counterclockwise loops both oppose $\ddot{B}_z^{(+2)}$. In other words, through this series of events, nature tries to reduce the original change of B made by the external agent. This type of 'self-compensating' effect is referred to as Lenz's law. Faraday's law is a Lenz law in electrodynamics. Notice that this self-compensating effect is caused by the negative sign on the right-hand side of equation (3.77). If there was no negative sign here, the direction of induced electric field $E^{(2)}$ would be

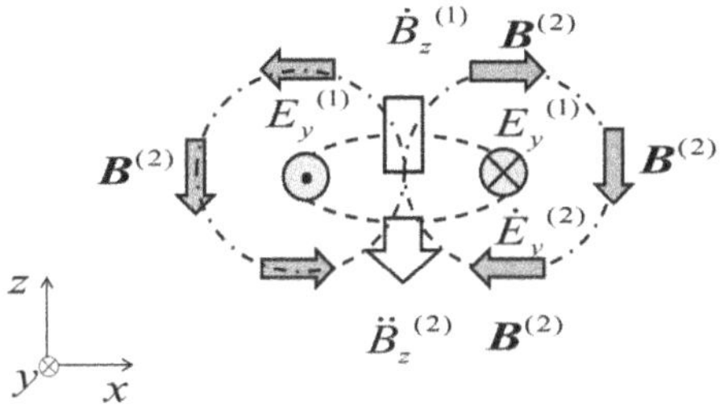

Figure 3.8. Pictorial explanation of generation of electromagnetic wave.

opposite to figure 3.8, which would enhance the initial change of magnetic field through the process discussed above.

Pictorial derivation of electromagnetic wave equation. Using figure 3.8 and the above discussion, we can argue the wave dynamics semi-quantitatively. For simplicity, we restrict the argument on the x-axis. However, the argument made here is applicable to any components of the field vectors.

Apply equation (3.77) to the induction on the x-axis as follows

$$\frac{\partial B_z}{\partial t} = -(\nabla \times E)_z = -\left(\frac{\partial E_y}{\partial x} - \frac{\partial E_x}{\partial y}\right) = -\frac{\partial E_y}{\partial x}. \tag{3.80}$$

In equation (3.80), we set the second term in the $(\nabla \times E)_z$ expression to zero because $E_x = 0$ on the x-axis.

Next, differentiate (3.80) with respect to time and apply equation (3.78) on the right-hand side

$$\frac{\partial^2 B_z}{\partial t^2} = -\frac{\partial}{\partial x}\frac{\partial E_y}{\partial t} = -\frac{1}{\epsilon\mu}\frac{\partial}{\partial x}\left(\frac{\partial B_x}{\partial z} - \frac{\partial B_z}{\partial x}\right) = \frac{1}{\epsilon\mu}\frac{\partial^2 B_z}{\partial x^2}. \tag{3.81}$$

Here we omit the first term in the parenthesis because $B_x = 0$ on the x-axis in going through the last equal sign.

Equation (3.81) is the equation for a wave whose phase velocity is $1/\sqrt{\epsilon\mu}$. From equations (3.80) and (3.81), we can see that the synergetic interaction between the electric and magnetic fields generates the wave dynamics. The pattern is that at a given time step the temporal change of one field induces the rotation of the other, and in the next time step the temporal change of the induced field induces the rotation of the other field. This pattern repeats in the subsequent time steps, causing the oscillatory behavior of the fields to continue and spread over the space.

Figure 3.9 illustrates this pattern pictorially. Here each row corresponds to a time step in a sequential order from the top to the bottom. At each time step, the dashed loop indicates the inducing field and the solid loop indicates the induced field. Referring to this figure, see the following pattern for the first four time steps.

(1) An external agent increases the magnetic field at the reference point. This results in a temporal change in the magnetic field. Faraday's law (3.77) induces an electric field E_1 in the space around the reference point.

(2) The induction of the electric field in step (1) newly generates an electric field in the space around the reference point. This results in a temporal change in the electric field. Consequently, Ampère's law induces a magnetic field B_2 in the space around the newly generated electric field E_1.

(3) The induction of the magnetic field in step (2) newly generates a magnetic field in the space around the electric field induced in step (1). This results in a temporal change in the magnetic field. Consequently, Faraday's law induces an electric field E_3 in the space around the newly generated magnetic field B_2.

(4) The induction of the electric field in step (3) newly generates an electric field in the space around the magnetic field induced in step (2). This results in a

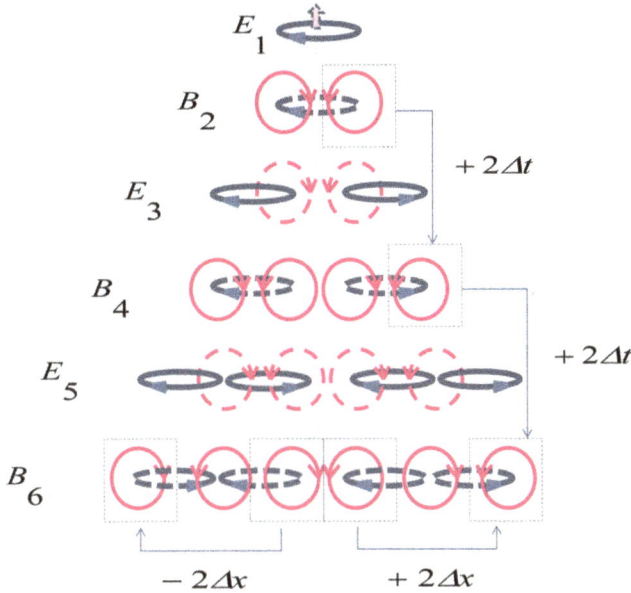

Figure 3.9. Pictorial explanation of generation of electromagnetic wave; several time steps. The dashed loop indicates the inducing field and the solid loop indicates the induced field. The labels on the left of the illustrations are the field induced at each time step.

temporal change in the electric field. Consequently, Ampère's law induces a magnetic field B_4 in the space around the newly generated electric field E_3.

In figure 3.9, pay attention to the pattern of the outermost induced magnetic field (enclosed by squares) on time steps 2, 4 and 6. We can find that moving forward in time by $2\Delta t$ and moving outward in space by $\pm 2\Delta x$ brings back the same pattern of the magnetic induction. By looking at odd time steps, we can see the same pattern in the induced electric fields. Mathematically, this means that the temporal secondary differentiation of the wave function is proportional to the spatial secondary differentiation of the wave function. This proportionality is exactly what the wave equation (3.81) represents.

Algebraic derivation of electromagnetic wave equation. We can derive the electromagnetic wave equation (3.81) from Faraday's law equation (3.77) and Ampère's law (3.78). As neither the electric charge nor current contributes to the generation of wave dynamics, consider the case where $\rho = j = 0$ on the right-hand side of equations (3.76) and (3.78). Note that the fact that the electric charge or current does not contribute to the generation of wave dynamics indicates that an electromagnetic wave can propagate in vacuum. This is in contrast to the case of sound waves where the particle velocity plays a role in generation of acoustic wave energy.

Take curl of equation (3.77) and use $\nabla \times E = \nabla\nabla \cdot E - \nabla^2 E$. From the condition $\rho = 0$, it follows that $\nabla \cdot E = 0$. Thus, we obtain the following equation

$$-\nabla^2 E = -\frac{\partial(\nabla \times B)}{\partial t}. \tag{3.82}$$

Now differentiate equation (3.78) with respect to time with the condition $j = 0$

$$\frac{\partial(\nabla \times B)}{\partial t} = \varepsilon\mu\frac{\partial^2(E)}{\partial t^2}. \tag{3.83}$$

From equations (3.82) and (3.83), we can eliminate B and find the following differential equation for E

$$\frac{\partial^2 E}{\partial t^2} = \frac{1}{\varepsilon\mu}\nabla^2 E. \tag{3.84}$$

Equation (3.84) is a wave equation in the same form as the wave equation (2.14) we derived in chapter 2. From comparison with equation (2.14), we find that the phase velocity in this case is

$$c = \frac{1}{\sqrt{\varepsilon\mu}}. \tag{3.85}$$

This phase velocity is known as the speed of light.

Eliminating E by taking curl of equation (3.78) and differentiating equation (3.78) with respect to time, we can derive a wave equation for B in the same form as E

$$\frac{\partial^2 B}{\partial t^2} = \frac{1}{\varepsilon\mu}\nabla^2 B. \tag{3.86}$$

There is some subtleness in Maxwell equations regarding the speed of light. Above we characterized the synergetic interaction between the electric and magnetic fields as the temporal differentiation of one with the spatial differentiation of the other. In section 2.1.5, we discussed the phase velocity of a wave as the ratio of the temporal periodicity to spatial periodicity. Now Faraday's law indicates the equivalence of temporal differentiation of B with spatial differentiation of E. These indicate that the ratio E/B can be the phase velocity of an EM wave, and this is true.

We can easily see that the ratio E/B is equal to the speed of light by using a sinusoidal magnetic wave $B_z = B_0 \sin(\omega t - kx)$ and analyzing E_y on the x-axis as we did for equation (3.80). Differentiate B_z in this form with respect to time

$$\frac{\partial B_z}{\partial t} = \omega B_0 \cos(\omega t - kx). \tag{3.87}$$

According to Faraday's law (3.77),

$$\frac{\partial B_z}{\partial t} = (\nabla \times E)_z = \left(\frac{\partial E_y}{\partial x} - \frac{\partial E_x}{\partial y}\right) = \frac{\partial E_y}{\partial x}. \tag{3.88}$$

Here $E_x = 0$ because we consider a wave traveling along the x-axis. Using equation (3.87) for the left-hand side of (3.88), it follows that we can express E_y as

$$E_y = \int \omega B_0 \cos(\omega t - kx) dx = \frac{\omega}{k} B_0 \sin(\omega t - kx) = \frac{\omega}{k} B_z. \qquad (3.89)$$

Equation (3.89) indicates that we can put E_y in the form of $E_y = E_0 \sin(\omega t - kx)$. As we discussed in chapter 2 with equation (2.41), ω/k is the phase velocity. Thus we find that the amplitude E_0 is related to B_0 as follows

$$E_0 = \frac{\omega}{k} B_0 = cB_0. \qquad (3.90)$$

If we start from Ampère's law (3.78) with $E_y = E_0 \sin(\omega t - kx)$ and repeat the same argument, we obtain the following

$$B_z = -\int (\omega \varepsilon \mu) E_0 \cos(\omega t - kx) dx = \frac{\omega}{k}(\varepsilon \mu) E_0 \sin(\omega t - kx) = \frac{1}{c} E_y. \qquad (3.91)$$

Here we use equation (3.85) for c in going through the last equal sign. Equation (3.91) is equivalent to equation (3.89). Although we used B_z and E_y above, the relation $E_0 = cB_0$ holds in general.

Boundary conditions for electromagnetic waves. The boundary conditions for an electromagnetic wave depend on the polarization; i.e., whether the electric field vector is parallel or normal to the plane of boundary. Expressing the components parallel and normal to the boundary surface with subscript p and n, we can write four boundary conditions for the $D = \varepsilon E$ and $B = \mu H$ fields as follows

$$(a)\ E_p^1 = E_p^2 \qquad (3.92)$$

$$(b)\ H_p^1 = H_p^2 \qquad (3.93)$$

$$(c)\ D_n^1 = D_n^2 \qquad (3.94)$$

$$(d)\ B_n^1 = B_n^2. \qquad (3.95)$$

We can understand these four boundary conditions by considering surface free electric charges σ_f and currents j_f on the boundary as illustrated in figures 3.10 and 3.11.

Consider first the charges inside a small block at the boundary in figure 3.10. The electric permittivity of the medium is ε_1 above the boundary and ε_2 below the boundary. By definition, the electric field is the spatial derivative of electric potential. Consider points A and B on the boundary. Suppose that the distance between the two points, δ, is infinitesimally small, and call the electric potential difference between the two points ϕ_e. The electric field is $E_p = -\nabla \phi_e = -\phi_e/\delta$. This field must be the same if you look at it below or above the boundary plane; $E_p^1 = E_p^2$. This is boundary condition (a).

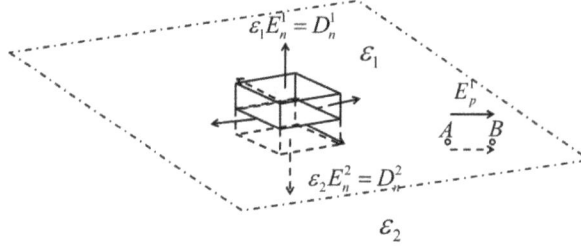

Figure 3.10. Surface electric charge and associated electric field on the boundary.

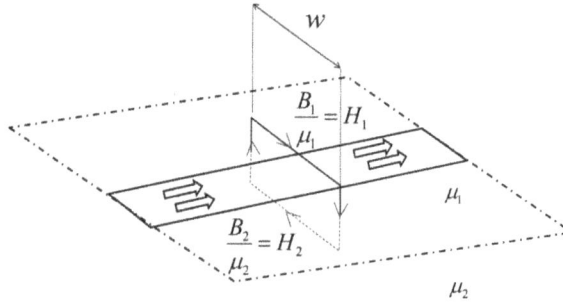

Figure 3.11. Surface electric current and associated magnetic field on the boundary.

According to Gauss's law equation (3.76) and the divergence theorem [8], we can express the electric field on the six surface planes of the block as follows

$$\oint (\varepsilon E) \cdot n \, dA = \int_{\text{top}} (\varepsilon_1 E) \cdot n \, dA + \int_{\text{bottom}} (\varepsilon_2 E) \cdot n \, dA = \sigma_f A. \qquad (3.96)$$

Here σ_f is the area density of the charge inside the block and A is the area of the block surface parallel to the boundary. The surface integration for the four side surfaces of the block (the surfaces perpendicular to the boundary) are zero because the height (thickness) of the block is zero; we can omit them in equation (3.96). Since the charges are uniformly distributed (because the block is infinitesimally small), we can replace the integration over the top and bottom surfaces with $(\varepsilon_1 E_n^1 + \varepsilon_2 E_n^2)A$. This leads to the following equation

$$\varepsilon_1 E_n^1 + \varepsilon_2 E_n^2 = \sigma_f. \qquad (3.97)$$

With the condition that there is no surface charge ($\sigma_f = 0$), and the linear medium condition ($\varepsilon_1 E_n^1 = D_n^1$ and $\varepsilon_1 E_n^2 = D_n^2$), equation (3.97) leads to condition (c).

We can derive boundary conditions (b) and (d) using Ampère's law equation (3.78) and the curl theorems as follows. Refer to figure 3.11 and consider the surface current flowing along the boundary. The magnetic permeability of the medium is μ_1 and μ_2 above and below the boundary

$$\oint \frac{B}{\mu} \cdot dl = \int_{\text{top}} \frac{B}{\mu_1} \cdot dl + \int_{\text{bottom}} \frac{B}{\mu_2} \cdot dl = j_f w. \qquad (3.98)$$

Here j_f is the linear current density (current per unit length perpendicular to the flow), and w is the width of the surface current across the flow. Again, we can set the line integrals for the left and right sides to zero because the height of the surface current is zero. Equation (3.98) leads to the following

$$\frac{B_p^1}{\mu_1} + \frac{B_p^2}{\mu_2} = j_f. \tag{3.99}$$

With no surface current $j_f = 0$ and the condition of linear medium ($B_p^1 = \mu_1 H_p^1$ and $B_p^2 = \mu_2 H_p^2$), equation (3.99) leads to condition (b).

We can derive condition (d) by considering $\nabla \cdot B = 0$ and the divergence theorem in a similar way to equation (3.96). It is obvious that $B_n^1 = B_n^2$; the normal component of B is continuous across the boundary.

For further details about the boundary conditions of the electric and magnetic fields, see chapters 4 and 6 of [9].

Reflection and transmission of electromagnetic waves. We can use boundary conditions (a) and (b) to find the reflection and transmission coefficients for electromagnetic waves. Since these boundary conditions are for the field components parallel to the boundary surface, we need to discuss the reflection and transmission depending on the polarization. Normally, the polarization is expressed in terms of the plane of incidence (figure 3.12). If the electric or magnetic field is perpendicular to the plane of incidence we use subscript $\perp$. If they are parallel to the plane of incidence, we use subscript $\parallel$. These parallel and perpendicular situations are different from the above subscripts p and n, which denote 'parallel' or 'normal' to the boundary plane, not the plane of incidence.

Reflection and transmission coefficients of $E_\perp$ wave. Referring to figure 3.13, consider the case of $E_\perp$. In this case, boundary conditions (a) and (b) read as follows

$$E_i + E_r = E_t \tag{3.100}$$

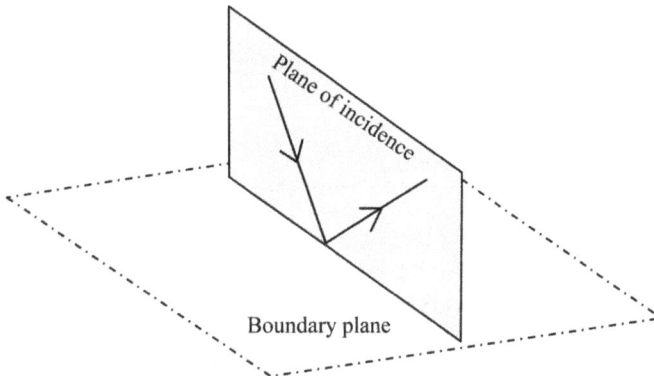

Figure 3.12. Plane of incidence.

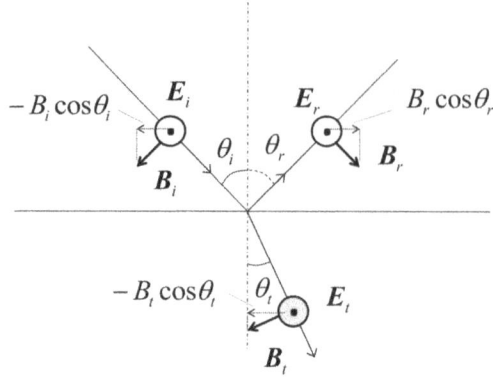

Figure 3.13. An electromagnetic wave incident to boundary plane with $E_\perp$ polarization.

$$-\frac{B_i}{\mu_1}\cos\theta_i + \frac{B_r}{\mu_1}\cos\theta_i = -\frac{B_t}{\mu_2}\cos\theta_t. \qquad (3.101)$$

Substituting relation $E_0 = cB_0$ (3.90) into equation (3.101) using the speed of light expression (3.85), we can rewrite the boundary condition as follows

$$-\sqrt{\frac{\varepsilon_1}{\mu_1}}\,E_i\cos\theta_i + \sqrt{\frac{\varepsilon_1}{\mu_1}}\,E_r\cos\theta_i = -\sqrt{\frac{\varepsilon_2}{\mu_2}}\,E_t\cos\theta_t. \qquad (3.102)$$

We can define a specific impedance for the electromagnetic wave as well. From equation (3.78), we can say H is directly related to the current density j and hence current-(velocity-)like. The electric field E is directly related to the electric force as $F_e = qE$, hence it is force-like. Thus, the impedance is E/H

$$z = \frac{E}{H} = \frac{E}{B}\mu = c\mu = \sqrt{\frac{\mu}{\varepsilon}}. \qquad (3.103)$$

From equations (3.100) (3.102) and (3.103), we obtain the reflection and transmission coefficient for $E_\perp$ case

$$R_\perp = \frac{\sqrt{\varepsilon_1/\mu_1}\cos\theta_i - \sqrt{\varepsilon_2/\mu_2}\cos\theta_t}{\sqrt{\varepsilon_1/\mu_1}\cos\theta_i + \sqrt{\varepsilon_2/\mu_2}\cos\theta_t} = \frac{z_2\cos\theta_i - z_1\cos\theta_t}{z_2\cos\theta_i + z_1\cos\theta_t} \qquad (3.104)$$

$$T_\perp = \frac{2\sqrt{\varepsilon_1/\mu_1}\cos\theta_i}{\sqrt{\varepsilon_1/\mu_1}\cos\theta_i + \sqrt{\varepsilon_2/\mu_2}\cos\theta_t} = \frac{2z_2\cos\theta_i}{z_2\cos\theta_i + z_1\cos\theta_t}. \qquad (3.105)$$

Equations (3.104) and (3.105) are known as Fresnel equations for s-polarized or TE waves. Here the polarization is for the electric field, and TE stands for transverse electric. The s of s-polarization comes from a German word *senkrecht* which means perpendicular. The polarization parallel to the plane of incidence is called the p-polarization (p stands for parallel).

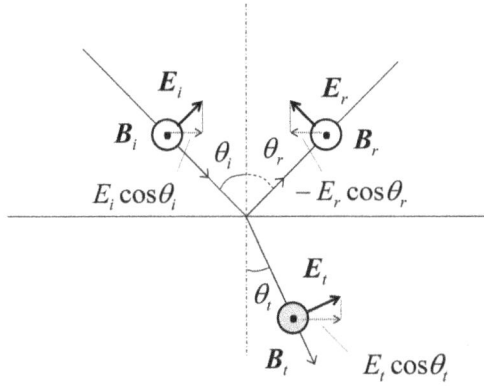

Figure 3.14. An electromagnetic wave incident to boundary plane with $E_\parallel$ polarization.

Reflection and transmission coefficients of $E_\parallel$ wave.   When the electric field is parallel to the plane of incidence, the boundary conditions are as follows; see figure 3.14

$$E_i \cos \theta_i - E_r \cos \theta_i = E_t \cos \theta_t \tag{3.106}$$

$$\frac{B_i}{\mu_1} + \frac{B_r}{\mu_1} = \frac{B_t}{\mu_2}. \tag{3.107}$$

Repeating the same argument as for the $E_\perp$ case, we can derive the following expressions for the reflection and transmission coefficient of the $E_\parallel$ case.

$$R_\parallel = \frac{\sqrt{\varepsilon_2/\mu_2}\,\cos \theta_i - \sqrt{\varepsilon_1/\mu_1}\,\cos \theta_t}{\sqrt{\varepsilon_2/\mu_2}\,\cos \theta_i + \sqrt{\varepsilon_1/\mu_1}\,\cos \theta_t} = \frac{z_1 \cos \theta_i - z_2 \cos \theta_t}{z_1 \cos \theta_i + z_2 \cos \theta_t} \tag{3.108}$$

$$T_\parallel = \frac{2\sqrt{\varepsilon_1/\mu_1}\,\cos \theta_i}{\sqrt{\varepsilon_1/\mu_1}\,\cos \theta_t + \sqrt{\varepsilon_2/\mu_2}\,\cos \theta_i} = \frac{2z_2 \cos \theta_i}{z_2 \cos \theta_t + z_1 \cos \theta_i}. \tag{3.109}$$

Equations (3.108) and (3.109) are known as Fresnel equations for p-polarized or TM waves. Here the polarization is for the electric field as is the case of TE waves. TM stands for transverse magnetic; if the polarization of E is p, the polarization of B is s or transverse.

3.2 Dispersion

As compared with reflection and refraction, dispersion may be somewhat of an abstract concept. In short, dispersion can be defined as the situation where the phase velocity depends on frequency. Since the phase velocity is the product of frequency and wavelength, we can rephrase it as 'wavelength (wavenumber) depends on the frequency', or 'frequency depends on the wavenumber'. An important consequence of these dependences is that the wave velocity is not a medium constant any more. This type of wave velocity is referred to as the group velocity, and will be discussed in the next chapter (section 4.1). Here, we explore the mechanism that makes the phase velocity nonconstant.

Similar to the earlier chapters, we first consider dispersion using a wave on a string. First, consider the origin of the phase velocity by rewriting the string wave equation (1.81) using equation (2.36)

$$k^2\frac{d^2\xi_y}{dt^2} = \omega^2\frac{d^2\xi_y}{dx^2}. \tag{3.110}$$

Apparently, the factor k^2 comes from the secondary spatial differentiation of the wave solution (1.82) and ω^2 from the secondary temporal differentiation. Equation (3.110) indicates that if the wave equation takes the form of 'secondary derivative of wave function with respect to time = constant × secondary derivative with respect to space', the phase velocity is the square of that constant, and hence it is a constant. If either the space or time derivatives have other orders of differentiation, this condition breaks and the phase velocity is no longer a constant.

3.2.1 String on an elastic medium

As an example, consider that a string is placed on an elastic medium of stiffness K, and that the elastic medium exerts vertical elastic force on the unit length of the string. This action by the elastic medium adds an additional term on the right-hand side of equation (1.81) as follows [10]

$$\frac{d^2\xi_y}{dt^2} = \frac{Tl}{m}\frac{d^2\xi_y}{dx^2} - \frac{K}{m}\xi_y. \tag{3.111}$$

The characteristic of equation (3.111) is given as follows

$$\omega^2 = \frac{Tl}{m}k^2 - \frac{K}{m}. \tag{3.112}$$

From equation (3.112) we find that the phase velocity $v_p \equiv \omega/k$ has the following form

$$v_p^2 = \left(\frac{\omega}{k}\right)^2 = \frac{Tl}{m} - \frac{K}{m}\frac{1}{k^2}. \tag{3.113}$$

Alternatively, from $v_p = \omega/k$, we can write equation (3.113) as follows

$$v_p^2 = \frac{Tl}{m} - \frac{K}{m}\frac{v_p^2}{\omega^2}.$$

Hence,

$$v_p = \sqrt{Tl\left(\frac{\omega^2}{m\omega^2 + K}\right)}. \tag{3.114}$$

Expression (3.114) explicitly indicates that the phase velocity varies as a function of frequency.

3.2.2 Decaying wave

In section 2.2.1, we observed that addition of the first-order time derivative term to the nondecaying plane wave equation makes the wave solution decay. (equations (2.12) and (2.21)). The angular frequency expression derived from the characteristic equation was as follows

$$\omega_0 = \sqrt{\frac{E}{\rho}k^2 - \beta^2}. \tag{2.50}$$

From equation (2.50), we obtain the following expression for the wave velocity.

$$v_p^2 = \left(\frac{\omega_0}{k}\right)^2 = \frac{E}{\rho} - \frac{\beta^2}{k^2} = \frac{E}{\rho} - \beta^2\frac{v_p^2}{\omega_0^2}. \tag{3.115}$$

Similar to the 'string on an elastic medium' case, we can find the frequency dependent wave velocity as follows

$$v_p = \sqrt{\frac{E}{\rho}\left(\frac{\omega_0^2}{\omega_0^2 + \beta^2}\right)}. \tag{3.116}$$

The decrease in the wave velocity with an increase in β corresponds to the decrease in the natural frequency of unforced oscillation due to β we observed in figure 1.4.

3.2.3 Light in matter

The phenomenon that white light passing through a triangular prism is separated into rainbow colors is well known as dispersion of light. As we discussed in section 3.1 (equation (3.1)), the light velocity in a medium is given by the speed of light in a vacuum c_0 and the refractive index of the medium n as c_0/n. Since c_0 is a constant, it follows that we can discuss the dispersion of light by considering the frequency dependence of the refractive index. The velocity of light is altered in a medium through the interaction between the electric field of the light wave and the electric charges of the medium. This interaction is known as the dielectric polarization. So, we start the discussion with a short description of dielectric polarization.

When an electric field is applied to a dielectric material, pairs of positive and negative charges are lined up as shown in figure 3.15. In this figure, the external field provided by the voltage source is applied to the conductive electrodes. Part of the gap between the electrodes is filled with a dielectric medium. A number of positive and negative charge pairs are formed inside the dielectric medium because the external electric field repels the positive charges and attracts the negative charges. Compare the electric fields inside the dielectric medium and outside. Obviously, the electric field inside the dielectric medium is smaller because the charge pairs

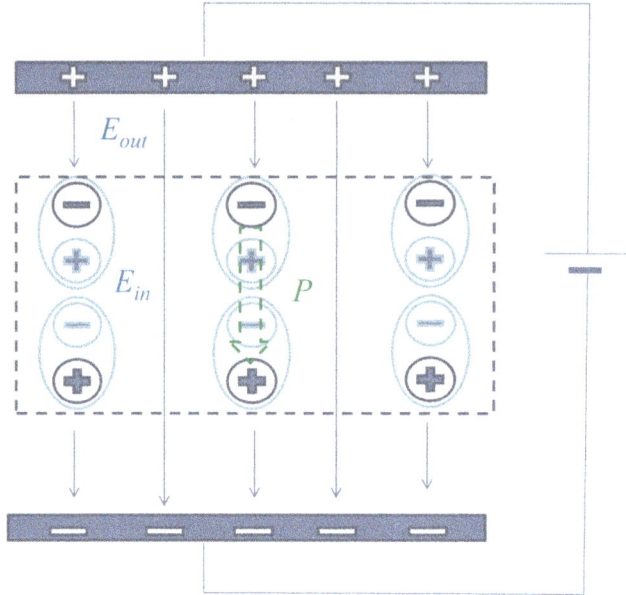

Figure 3.15. Polarization vector in dielectric medium.

neutralize the external field. We can describe the situation with the following equation

$$D = \varepsilon_0 E + P. \tag{3.117}$$

Here D is the electric displacement (or flux density) vector, ε_0 is the electric permittivity in vacuum, and P is the polarization vector. With this definition, we can view P as the quantity that weakens the effect of the applied electric field; a natural reaction of the medium to reduce the external electric field.

In a linear medium, D is proportional to the electric field and the constant of the proportionality ε is known as the permittivity of the medium

$$D = \varepsilon E \tag{3.118}$$

$$\varepsilon_r = \frac{\varepsilon}{\varepsilon_0} > 1. \tag{3.119}$$

Here ε_r is called the dielectric constant. In figure 3.15, notice that E is not continuous through the dielectric field but D is. We can view that the greater permittivity inside the dielectric medium compensates the smaller electric field so that the electric displacement vector is continuous. We can say that D is directly related to the charge while E is directly related to the electric potential.

From the electrodynamic point of view, we can interpret the polarization vector as the collective effect of the dipole moment of electrons. When a dielectric medium is placed in the oscillatory electric field of an electromagnetic wave, the electric field exerts electric force on the electron orbiting around the nucleus. We can model the dynamics with the damped driven harmonic oscillation we discussed in chapter 1 (section 1.2.3 and equation (1.42)). We can write the equation of motion for the electron as follows

$$m\frac{d^2\tilde{x}}{dt^2} + b_{ef}\frac{d\tilde{x}}{dt} + k_{ef}\tilde{x} = qE_0e^{i\omega t}. \tag{3.120}$$

Here $\tilde{x}$ is the complex displacement of the electron, m and q are the mass and charge of the electron, b_{ef} is the effective coefficient for the velocity damping force representing the energy dissipative nature of the electron's motion, k_{ef} is the effective spring constant representing the oscillatory nature of the electron motion, E_0 and ω are the amplitude and angular frequency of the electric field of the electromagnetic wave. The displacement $\tilde{x}$ is a complex number because of the phase delay from the driving electric field. By dividing both sides by m and putting the other two coefficients on the left-hand side as $b_{ef}/m \equiv \gamma$ and $k_{sp}/m \equiv \omega_0^2$ (as we did in section 1.2.3), we can rewrite equation (3.120) in the following form

$$\frac{d^2\tilde{x}}{dt^2} + \gamma\frac{d\tilde{x}}{dt} + \omega_0^2\tilde{x} = \frac{qE_0}{m}e^{i\omega t}. \tag{3.121}$$

The dipole moment $\tilde{p}$ is defined as the product of the electric charge q and the oscillation amplitude $\tilde{x}$. Since the charge q is real and $\tilde{x}$ is complex, $\tilde{p}$ is complex.

We know that the solution to differential equation (3.121) can take the following form

$$\tilde{x}(t) = \frac{q/m}{\omega_0^2 - \omega^2 - i\gamma\omega}E_0e^{i\omega t}. \tag{3.122}$$

Thus the dipole moment is

$$\tilde{p}(t) = q\tilde{x} = \frac{q^2/m}{\omega_0^2 - \omega^2 - i\gamma\omega}E_0e^{i\omega t}. \tag{3.123}$$

In each molecule, there are a number of electrons having different natural frequencies and damping coefficients. Identifying these electrons with suffix j and expressing the total number of molecules in a unit volume with N, we can express polarization P as the overall effect

$$\tilde{P} = \frac{Nq^2}{m}\left(\sum_j \frac{f_j}{\omega_{j0}^2 - \omega^2 - i\gamma_j\omega}\right)\tilde{E}. \tag{3.124}$$

Here f_j is the number of electrons that have natural angular frequency ω_{j0} and damping coefficient γ_j.

Now we define the electric susceptibility $\tilde{\chi}_e$ as follows

$$\tilde{P} = \varepsilon_0\tilde{\chi}_e\tilde{E}. \tag{3.125}$$

Comparing equations (3.124) and (3.125) we find

$$\tilde{\chi}_e = \frac{Nq^2}{m\varepsilon_0}\left(\sum_j \frac{f_j}{(\omega_{j0}^2 - \omega^2) - i\gamma_j\omega}\right). \tag{3.126}$$

By substituting equation (3.125) into equation (3.117), we obtain the following expression

$$\tilde{D} = \varepsilon_0\tilde{E} + \varepsilon_0\tilde{\chi}_e\tilde{E} = \varepsilon_0(1 + \tilde{\chi}_e)\tilde{E}. \tag{3.127}$$

Comparison of equations (3.118) and (3.127) leads to the following relation between the electric permittivity and susceptibility

$$\tilde{\varepsilon}_r = \frac{\tilde{\varepsilon}}{\varepsilon_0} = (1 + \tilde{\chi}_e). \tag{3.128}$$

Here we applied the dielectric constant expression (3.119) to the complex electric permittivity.

On the other hand, from the discussion in chapter 2 (see section 2.1.4), we know that the wave equation for the electric field of an electromagnetic wave propagating through an energy dissipative medium can take the following form [11]

$$\frac{\partial^2\tilde{E}}{\partial t^2} - \frac{1}{\tilde{\varepsilon}\mu}\nabla^2\tilde{E} = 0. \tag{3.129}$$

Here we combine the oscillatory and energy dissipative effects corresponding to E/ρ (the square of the mechanical wave velocity) and β (decay constant) in equation (2.21) into the complex electromagnetic version of them $1/(\tilde{\varepsilon}\mu)$; the damping effect is absorbed in the imaginary part of $\tilde{\varepsilon}$.

As we discussed for equation (2.21), a plane wave solution is

$$\tilde{E}(z, t) = \tilde{E}_0 e^{i(\omega t - \tilde{k}z)} \tag{3.130}$$

where $\tilde{k}$ is the complex wave number

$$\tilde{k} = k_0 + ik_i. \tag{3.131}$$

Its imaginary part represents the energy dissipative nature of the medium. Substitution of solution (3.130) into wave equation (3.129) leads to the following expression of $\tilde{k}$

$$\begin{aligned}
\tilde{k} &= \sqrt{\tilde{\varepsilon}\mu_0}\,\omega = \sqrt{\tilde{\varepsilon}_r\varepsilon_0\mu_0}\,\omega = \sqrt{\tilde{\varepsilon}_r}\frac{\omega}{c_0}\\[4pt]
&\cong \frac{\omega}{c_0}\left(1 + \frac{1}{2}\tilde{\chi}_e\right)\\[4pt]
&\cong \frac{\omega}{c_0}\left[1 + \frac{Nq^2}{2m\varepsilon_0}\left(\sum_j \frac{f_j}{(\omega_{j0}^2 - \omega^2) - i\gamma_j\omega}\right)\right].
\end{aligned} \tag{3.132}$$

Here we used $\varepsilon_r = 1 + \tilde{\chi}_e$ (equation (3.128)) using equation (3.126) for $\tilde{\chi}_e$, and binomial approximation $\sqrt{1 + \tilde{\chi}_e} \approx 1 + \tilde{\chi}_e/2$ assuming $1 >> \tilde{\chi}_e$.

Separating the right-hand side of equation (3.132) into the real and imaginary part, we can clarify the physical meanings of n as the phase part of the interaction between the driving electric field (light) and the dipole

$$n = \frac{c_0}{c} = \frac{c_0}{\omega/k_0} = 1 + \frac{Nq^2}{2m\varepsilon_0}\left(\sum_j \frac{f_j\left(\omega_{j0}^2 - \omega^2\right)}{\left(\omega_{j0}^2 - \omega^2\right)^2 + \gamma_j^2\omega^2} \right) \qquad (3.133)$$

$$k_i = \frac{\alpha}{2} = \frac{Nq^2}{2m\varepsilon_0 c_0}\left(\sum_j \frac{f_j\gamma_j\omega^2}{\left(\omega_{j0}^2 - \omega^2\right)^2 + \gamma_j^2\omega^2} \right). \qquad (3.134)$$

Here α is the power absorption coefficient. The factor 2 comes from the fact that the power P is proportional to E^2 hence $P \propto (e^{-k_i z})^2 = e^{-2k_i z} = e^{-\alpha z}$.

Figure 3.16 plots $n - 1$ and α as a function of optical frequency. Note that the frequency dependence of the index of refraction is very strong in the near resonance frequency range but it reduces fairly sharply as the frequency deviates from the resonance in either direction. This means that as far as the frequency of the incident light is outside this near resonance range the effect of dispersion is low but it can rapidly increase as the frequency shifts toward the resonance. This is good to know when you design an optical system using transmissive optics such as lenses and polarizers. We may encounter the situation where the frequency dependence of refractive index is negligibly small for normal usage of the system but it has a considerable effect when the light frequency is shifted toward the resonance. As an example, you may have a situation where a prism-type polarizer suddenly starts to perform poorly and the problem is not in the polarizer but the change in the frequency of the incident light.

The α-curve in figure 3.16 indicates that as an oscillatory system, the molecule can absorb the electromagnetic energy of the incoming light only when its frequency is

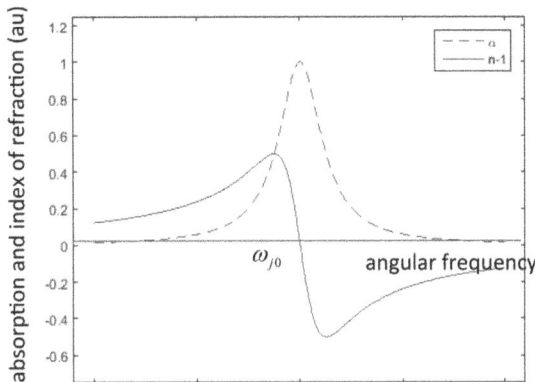

Figure 3.16. Frequency response of dipole to incoming light's electric field.

within the near resonance range. This observation naively explains the wave–particle duality of light. The photon energy is known to be $\hbar\omega$. In the context of figure 3.16, this ω means the resonant frequency. The absorption takes place only for the near-resonance frequency component of the incoming electromagnetic field. It is similar to the fact that a string of a fixed length oscillates at only discrete values of frequency, as we discussed at the end of chapter 1. In the same way that the string oscillates at an integer multiple of the fundamental frequency $\sqrt{T/(ml)}/2$ (equation (1.87)), the molecular system absorbs an integer multiple of unit energy $\hbar\omega$. It should be noted that this discreteness comes from the fact that the molecular system has discrete energy levels; it does not come from the nature of the incoming light.

This view of light–matter interaction explains the stimulated emission to some extent as well. The molecular system exchanges energy with an oscillatory electric field only through this small window of frequency. The wave function that describes the molecular state and associated energy is a solution to the corresponding Schrödinger's wave equation. In a sense, this light–matter interaction is a wave-versus-wave interaction.

3.2.4 Transverse wave in beam

Similar to the case of waves on a string, transverse waves can be excited in a steel beam. Consider that a beam of uniform cross-sectional area A is being bent. Figure 3.17 illustrates a small segment of the beam. Here the coordinate axis s is along a neutral line whose radius of curvature is R. Due to the bending action, the steel above the neutral line is stretched, and the steel below the neutral line is compressed. Expressing the stretch at radius r above the neutral line with δs, we can relate the stretch to the force acting on the right end of the segment exerted by the neighboring segment df

$$df = E\,dA\frac{\delta s}{ds}. \tag{3.135}$$

Here ds is the arc length before the stretch, and we can view the stretch δs as the difference in arc length at radii $R + r$ and R. E is the Young's modulus and the fraction $\delta s/ds$ represents strain. Equation (3.135) is equivalent to equation (2.10) that we discussed for a longitudinal wave in chapter 2.

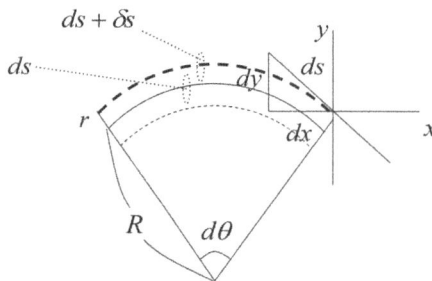

Figure 3.17. Bent steel beam.

$$\frac{ds + \delta s}{R + r} = \frac{ds}{R}. \tag{3.136}$$

From equation (3.136) we find

$$\frac{ds + \delta s}{ds} = \frac{R + r}{R}. \tag{3.137}$$

and hence

$$\frac{\delta s}{ds} = \frac{r}{R}. \tag{3.138}$$

Substituting equation (3.138) we can rewrite equation (3.135) as follows

$$df = E \, dA \frac{r}{R}. \tag{3.139}$$

Using equation (3.139) we can write the bending moment M in the following form

$$M = \int r df = \frac{E}{R} \int r^2 \, dA = \frac{E}{R} A \eta^2. \tag{3.140}$$

Here E is Young's modulus and

$$\eta^2 = \frac{\int r^2 \, dA}{A}. \tag{3.141}$$

is a constant determined by the shape of the beam.

Take x- and y-axes horizontally and vertically as shown in figure 3.18. The transverse wave is excited by the differential bending moment at x. Before considering the differential bending moment, we need to express the radius of curvature with x and y. Since ds is the arc length subtended by angle $d\theta$ and hence $ds = R \, d\theta$, we find the following equality through differentiation with respect to x

$$\frac{ds}{dx} = R \frac{d\theta}{dx}. \tag{3.142}$$

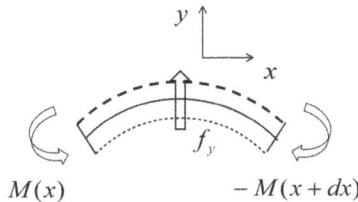

Figure 3.18. Differential moment.

On the other hand,

$$\tan \theta = \frac{dy}{dx}. \tag{3.143}$$

Differentiate both sides of equation (3.143) with respect to x

$$\frac{d}{dx} \tan \theta = \frac{d}{d\theta} \tan \theta \frac{d\theta}{dx} = (1 + \tan^2 \theta) \frac{d\theta}{dx}$$

$$= \left(1 + \left(\frac{dy}{dx}\right)^2\right) \frac{d\theta}{dx} \equiv (1 + y'^2) \frac{d\theta}{dx} \tag{3.144}$$

$$\frac{d}{dx} \frac{dy}{dx} = \frac{d^2 y}{dx^2} \equiv y''. \tag{3.145}$$

From equations (3.143), (3.144) and (3.145), we find

$$\frac{d\theta}{dx} = \frac{y''}{1 + y'^2}. \tag{3.146}$$

Since we can express ds in terms of dx and dy as $ds^2 = dx^2 + dy^2$

$$\left(\frac{ds}{dx}\right)^2 = 1 + \left(\frac{dy}{dx}\right)^2. \tag{3.147}$$

From equations (3.142), (3.146) and (3.147), we find

$$R = \frac{ds/dx}{d\theta/dx} = \frac{[1 + y'^2]^{3/2}}{y''}. \tag{3.148}$$

Under the condition $\partial y / \partial x \ll 1$, we can neglect the square term in the numerator in equation (3.148), and make the following approximation

$$R \approx \frac{1}{y''}. \tag{3.149}$$

With equation (3.149), we can express the bending moment (3.140) as follows

$$M = \frac{E}{R} A \eta^2 \approx E A \eta^2 \frac{\partial^2 y}{\partial x^2}. \tag{3.150}$$

Now consider the bending moments on the left and right edge of the segment ds in figure 3.18. Taking the counterclockwise moment as positive (because the x-axis is rightward and y-axis is upward), we can interpret that the differential moment $\partial M / \partial x$ is due to the vertical upward force $-f_y(x)$. Using equation (3.150), we obtain the following expression

$$f_y = -\frac{\partial M}{\partial x} = -E A \eta^2 \frac{\partial^3 y}{\partial x^3}. \tag{3.151}$$

Since the transverse wave is caused by the differential force $df_y = (\partial f_y/\partial x)dx$, we obtain the following equation of motion

$$\rho A\, dx\frac{\partial^2 y}{\partial t^2} = -EA\eta^2\frac{\partial^4 y}{\partial x^4}dx. \qquad (3.152)$$

As before, the equation of motion (3.152) leads to the following wave equation

$$\frac{\partial^2 y}{\partial t^2} = -\frac{E\eta^2}{\rho}\frac{\partial^4 y}{\partial x^4} = -\eta^2 v_{pl}^2\frac{\partial^4 y}{\partial x^4}. \qquad (3.153)$$

Here $v_{pl} = \sqrt{E/\rho}$ is the phase velocity of longitudinal waves.

Apparently, the $y = f(\omega t - kx)$ type of solution does not work for the equation of motion (3.153) because of the fourth order spatial differentiation on the right-hand side. Assuming exponential time dependence, we can try separation of variables

$$y = \Psi(x)e^{i\omega t}. \qquad (3.154)$$

Substitution of equation (3.154) into equation (3.152) leads to the following differential equation for the space function

$$\frac{\partial^4 \Psi}{\partial x^4} = \frac{\omega^2}{\eta^2 v_{pl}^2}\Psi. \qquad (3.155)$$

As usual, try a test function in the form of $\Psi(x) = \Psi_0 e^{\gamma x}$ and substitute it into equation (3.155). We obtain the following expression for the wave number γ

$$\gamma^4 = \frac{\omega^2}{\eta^2 v_{pl}^2}. \qquad (3.156)$$

Now viewing γ as being related to the phase velocity of the transverse wave v_{pt} as $v_{pt} = \omega/\gamma$, we can obtain the following form for v_{pt}

$$\gamma^4 = \frac{\omega^4}{v_{pt}^4}. \qquad (3.157)$$

Comparison of equations (3.156) and (3.157) tells us that the phase velocity of the transverse wave has the following form

$$v_{pt} = \sqrt{\omega\eta v_{pl}}. \qquad (3.158)$$

Equation (3.158) shows the frequency dependence of the transverse wave velocity.

Equation (3.157) yields four possibilities for the wave number γ, which in turn leads to the general solution of the transverse wave in the following form

$$y = e^{i\omega t}(Ae^{\omega x/v_{pt}} + Be^{-\omega x/v_{pt}} + Ce^{i\omega x/v_{pt}} + De^{-i\omega x/v_{pt}}). \qquad (3.159)$$

We can determine coefficients A–D by boundary conditions.

3.2.5 Beads on a string

As the final example of dispersion, we briefly discuss waves propagating on a classical system known as beads on a string [12]. As usual, we start from the equation of motion. However, derivation of the wave equation is somewhat lengthy and will be omitted here. We will focus on the dispersion of the wave and discuss some of the physics behind it.

A system of beads on a string is defined as a number of beads connected with massless strings of the same length. Consider the transverse motion of the beads in figure 3.19 where beads of mass m are connected with strings of length l. We can express the equation of motion governing the transverse motion of three neighboring beads (labeled $n-1$, n and $n+1$) as follows

$$m\frac{\partial^2 \xi}{\partial t^2} = -T \sin\theta_1 + T \sin\theta_2 = -T\left(\frac{\xi_n - \xi_{n-1}}{l}\right) + T\left(\frac{\xi_{n+1} - \xi_n}{l}\right). \tag{3.160}$$

Here T is the tension, which is common to all beads, and θ_1 are the angles made by the strings connecting mass $n-1$ and n and the horizontal axis (the x-axis). θ_2 is the angle between the string connecting n and $n+1$ and the x-axis.

For simplicity, rewrite equation (3.160) in the following form

$$\frac{\partial^2 \xi}{\partial t^2} = \omega_0^2 (\xi_{n+1} - 2\xi_n + \xi_{n-1}). \tag{3.161}$$

Here ω_0 is defined as follows

$$\omega_0^2 = \frac{T}{ml}. \tag{3.162}$$

We can derive a total of N equations in this form for the total number of mass N.

Substituting solutions in the form of sinusoidal time dependence into the N equations of motion and going through somewhat lengthy mathematical manipulation (mathematically lengthy but physically rather straightforward), we obtain the following expression for the displacement of nth mass

$$\xi_n = \sum_{j=1}^{N} A^{(j)} \sin\left(n\frac{j\pi}{N+1}\right) \cos(\omega_j t + \alpha_j), \, j = 1, 2, \cdots N. \tag{3.163}$$

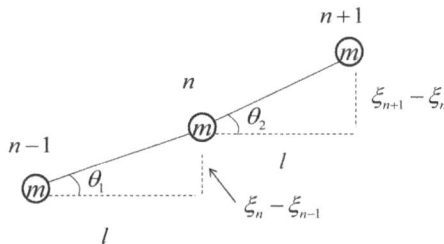

Figure 3.19. Beads on string.

Here we can determine $A^{(j)}$ and α_j from the initial conditions of the masses and express ω_j as follows

$$\omega_j = 2\omega_0 \sin \frac{j\pi}{2(N+1)}, j = 1, 2, \cdots N. \tag{3.164}$$

By expressing the total length of the string with L, we can interpret the discrete value of $n/(n+1)$ in the argument of the sine function in equation (3.163) as a continuous quantity x/L. Thus, we can rewrite the sine function as follows

$$A^{(j)}(x) = A^{(j)}\sin \frac{j\pi}{L}x. \tag{3.165}$$

Further, by viewing $L/(2j)$ as the wavelength λ of the continuous function, we can rewrite equation (3.165) with wavenumber $k = 2\pi/\lambda = j\pi/L$ as follows

$$A^{(j)}(x) = A^{(j)}\sin \frac{2\pi}{\lambda}x = A^{(j)}\sin kx. \tag{3.166}$$

Similarly, the discrete angular frequency becomes

$$\omega = 2\omega_0 \sin \left(\frac{kL}{2(N+1)}\right) = 2\omega_0 \sin \left(\frac{kl}{2}\right) \tag{3.167}$$

hence,

$$v_p(k) = \frac{\omega}{k} = \frac{2\omega_0 \sin(kl/2)}{k}. \tag{3.168}$$

Equation (3.168) indicates the dispersion of the wave.

In the limit where $l \ll 1$ so that we can view the beads on a string as a continuous string with mass, the phase velocity reduces to the one we found for the wave on a string in section 2.1.5 (see equation (2.36))

$$v_p = (k) = \frac{2\omega_0 \sin(kl/2)}{k} \approx \frac{2\omega_0 (kl/2)}{k} = \omega_0 l = l\sqrt{\frac{T}{ml}} = \sqrt{\frac{T}{m/l}}. \tag{3.169}$$

We can interpret the dispersion of this case as follows. As we discussed previously, the wave is the continuous propagation of oscillation where each segment of the continuum oscillates at its natural frequency. When the massless string has N masses, there are N modes each having its fundamental frequency. That is why the discrete angular frequency is a function of l. In this situation, the system allows only those discrete frequencies for continuous oscillation. In other words, waves of these frequencies can only propagate the bead on string system. When the string becomes continuum with mass, the frequency separation becomes null (meaning that the frequency becomes continuous), and the string can propagate at any frequency.

Thus, we can say that the dispersion is due to the frequency selectivity of the system. Similar dispersion occurs in waveguides [13].

3.3 Interference

In chapter 2 we discussed the magnitude and phase of superposed waves. Figure 2.10 indicated that the magnitude of the superposed wave varies from a minimum to maximum value depending on the relative phase difference between the component waves. We say that the component waves interfere with each other and the resultant strength is determined by the relative phase difference. In chapter 2, we also discussed that we can add waves as vectors on the complex plane and that the result of addition can be represented by a pair of magnitude and phase. This means that no matter how many waves we may be dealing with, we can add one wave at a time and argue the resultant wave with a pair of magnitude and phase. So, although here we discuss interference of two waves, the argument is applicable to interference of multiple waves.

As I briefly mentioned in chapter 2, the interference can be used to detect the phase difference between interfering waves through analysis of the total intensity. This technique is called interferometry. Here we focus our attention on interference in the context of interferometers. As examples, we will discuss optical and acoustic interferometry.

Consider that a pair of waves (the component waves) are interfering with each other

$$
\begin{aligned}
I &= (A_1 e^{i\theta_1} + A_2 e^{i\theta_2})(A_1 e^{i\theta_1} + A_2 e^{i\theta_2})^* \\
&= A_1 A_1^* + A_2 A_2^* + e^{i(\theta_1 - \theta_2)} + e^{-i(\theta_1 - \theta_2)} \\
&= A_1^2 + A_2^2 + 2\cos(\theta_1 - \theta_2).
\end{aligned}
\tag{3.170}
$$

Here the subscripts 1 and 2 denote the component waves, respectively. Each individual phase can be put in the form of $\theta = \omega t - kz$. In principle, the relative phase can be due to a difference in time or space coordinate. If the second wave is launched Δt later than the first wave and the interference occurs at z, $\Delta\theta = \omega_1 t_1 - \omega_2(t_1 + \Delta t)$. Normally, however, interference occurs when two waves from the same source are superposed where one of the waves takes a longer path to reach the point of interference. So, it is natural to assume $\omega_1 = \omega_2 = \omega$ and $k_1 = k_2 = k$, and set $z_1 = z_2 + \Delta z$. In this context, $\Delta\theta = (\omega t + k(z_1 + \Delta z)) - (\omega t + kz_1) = k\Delta z$. With this configuration of interferometer, we can measure the path length difference.

In interferometry, it is important to have the amplitude of the component waves as close as possible to each other. Here we will see why. Figure 3.20 is the same type of figure as figure 2.10 where two cases (a) $A_1 = A_2$ and (b) $A_2/A_1 = 0.5$ are plotted for comparison. The vertical axis is intensity I (because in most experiments we measure intensity) as opposed to $\sqrt{I}$ in figure 2.10. Notice that (a) exhibits the higher maximum value and lower minimum value as compared with (b). We say that (a) has a higher contrast than (b). In interferometry, the contrast is measured by a parameter called the visibility. The visibility V is defined as follows

$$
V = \frac{I_{\max} - I_{\min}}{I_{\max} + I_{\min}}.
\tag{3.171}
$$

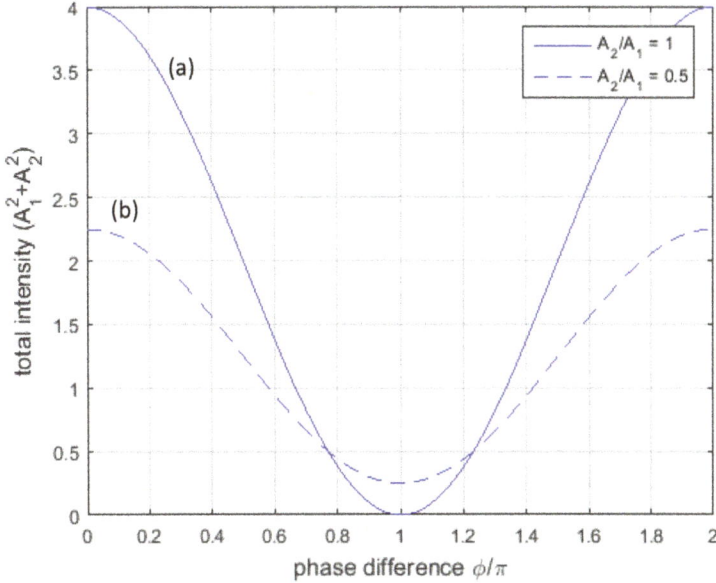

Figure 3.20. Two cases of interference. (a) $A_1 = A_2$ and (b) $A_2/A_1 = 0.5$.

Here $I_{\max}$ and $I_{\min}$ are the maximum and minimum intensity. In the ideal situation where the contrast is highest, $I_{\min} = 0$ and hence the visibility is unity. In the worst case, the high and low peaks of interference are similar to each other, and hence the visibility is close to zero. In the case of figure 3.20, the visibility of (a) is 1 and (b) is 0.8.

3.3.1 Optical interferometry

As an example of interferometer, I would like to briefly discuss a laser interferometric gravitational wave detector. My intention here is not to discuss details of gravitational waves or their sources. Instead, I would like to give an intuitive picture of the detection of gravitational waves with optical interferometry as it demonstrates the concept of phase based on profound physics. Those who are interested in more information about this subject are encouraged to read other references [14, 15]. Also, a book in this series focuses on gravitational waves.

A gravitational wave is a tiny fluctuation of the gravitational field strength that propagates at the speed of light. When a gravitational wave passes through the Earth, the gravitational field strength increases in one direction and decreases in an orthogonal direction. Imagine that you squeeze a tennis ball. It gets compressed in the direction of the applied compressive force and expanded orthogonally. The gravitational field is an acceleration field, so it is proportional to force. The gravity near the Earth behaves like this squeezing force when a gravitational wave passes through. Consequently, the space–time experiences tiny oscillatory motion. In terms of strain, the space–time fluctuation is typically of the order of 10^{-23} or smaller.

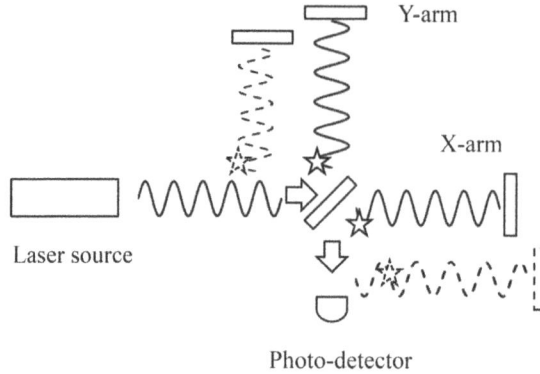

Figure 3.21. Optical interferometric gravitational wave detector.

A gravitational wave detector is designed to detect such tiny fluctuations in space–time with an extremely sensitive optical interferometer. Figure 3.21 illustrates the principle of operation schematically. The beam splitter splits the laser beam into two interferometric arms, called the X-arm and Y-arm. At the end of each arm, a total reflector reflects the laser light back to the beam splitter where the two returning light waves interfere with each other. Initially, the length of the X and Y arms are adjusted so that the interference at the beam splitter is totally destructive; there is no light coming out through the beam splitter toward the photo-detector. When a gravitational wave expands the space parallel to one arm and contracts the space parallel to the other arm, this destructive interference breaks and the photo-detector captures the light leaking out from the interferometer.

To understand how and why the above mechanism works, we need to understand the relation between gravitational field strength and space–time. Again, I am only giving a brief explanation here. Imagine that a laser light coming from a far distance is passing near the planet in figure 3.22. According to general relativity [16], the gravity of the planet bends the light beam. Consider Wavefront 1 and 2 over which there is one wave (the distance between the two wavefronts is equal to one wavelength). The laser beam is so wide that the inner edge and outer edge of the beam take different trajectories called Orbit 1 and Orbit 2. Being wavefronts, Wavefront 1 and 2 can be characterized as follows. At the moment when the inner edge of the laser beam is at point P_1^1 the outer edge of it is at P_1^2. The light near the inner edge and outer edge travel along Orbit 1 and Orbit 2, respectively, to reach Wavefront 2. Obviously, Orbit 2 is longer than Orbit 1. However, since Wavefront 2 is one wave away from Wavefront 1, if we measure the distances that the two portions of the light beam travel during this time in phase, both are 2π. It follows that the duration in time is longer along Orbit 2 than Orbit 1. (People may say 'the clock in Orbit 2 is faster than Orbit 1', but I think it is confusing to explain the situation using clocks.) Thus, we can characterize the two wavefronts as they are one period different from each other either in time or space.

We can discuss the above argument from slightly different viewpoints. First, remember that in early chapters we interpreted the phase velocity of a wave as the

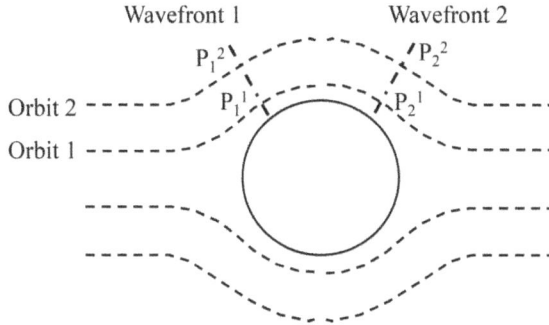

Figure 3.22. Gravity of planet bends light.

ratio of the temporal periodicity to the spatial periodicity. From this standpoint, we can describe the situation as follows. Although the portions of the laser beam near the inner edge and outer edge experience different temporal and spatial distances, they have a common phase velocity, i.e., the speed of light, as the ratio of the respective temporal to spatial periodicities, is the same along the two orbits. Second, the frequency is the reciprocal of the time. Along Orbit 1, the duration in time is shorter than Orbit 2 as much as the spatial duration is shorter. This indicates that the frequency along Orbit 1, which feels stronger gravity by the planet than Orbit 2, is higher. Photon energy is Planck constant times frequency. The photon energy in Orbit 1 is higher than Orbit 2. Einstein says that the photon receives energy from gravity [17].

Now go back to figure 3.21 and consider that a gravitational wave causes the gravity along the X-arm to be less than usual and along the Y-arm to be greater than usual. The situation here is equivalent to the two orbits in figure 3.22. The X-arm corresponds to Orbit 2 and the Y-arm corresponds to Orbit 1. According to the argument made above, the frequency along the Y-arm is higher than the X-arm. Therefore, when the light traveling along the Y-axis returns to the beam splitter, the light traveling along the X-arm experiences fewer wave numbers than the Y-arm. In figure 3.21, the dashed star marks near the beam splitter indicate the number of waves that the two laser beams experience when returning to the beam splitter. Initially, at the beam splitter, the relative phase of the laser beams along the two arms are set to be destructive, as indicated by the solid star marks in figure 3.21. Apparently, the two dashed star marks do not cause destructive interference. This is how this type of optical interferometric gravitational-wave detector detects a signal.

3.3.2 Acoustic interferometry

As an example of acoustic interferometry, I would like to make a short description of scanning acoustic microscopy (SAM). SAM is used for characterization of the elastic property near the surface of solid specimens. Figure 3.23 illustrates the principle of operation. A typical SAM system consists of an acoustic transducer, acoustic lens and coupling medium. The acoustic transducer placed on the top surface of the lens sends out an acoustic wave at a certain range of frequency toward

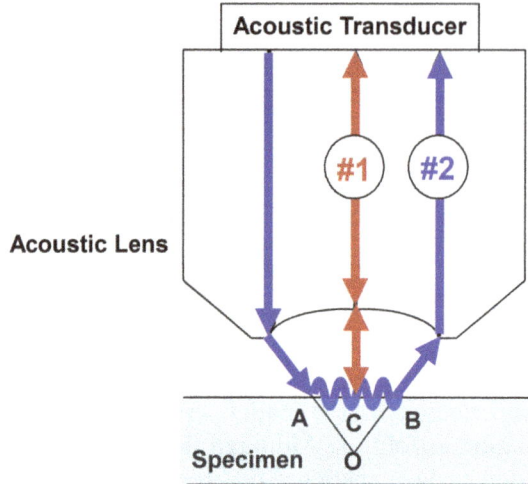

Figure 3.23. Scanning acoustic microscope.

the specimen placed under the lens through the coupling medium (liquid, usually distilled water, not shown in figure 3.23). The acoustic lens is designed to apply the acoustic wave to the specimen at a certain angle of incidence so that the so-called leaky surface acoustic wave is generated and propagates along the liquid–solid interface. The surface wave generates oscillatory motion at the liquid–solid interface, which radiates energy back into the liquid.

The wave due to this acoustic radiation interferes with the specularly reflected wave at the top of the lens. The acoustic transducer detects the overall (superposed) wave and outputs as a voltage signal. The two acoustic paths labeled #1 and #2 in figure 3.23 represent, respectively, the acoustic wave that reaches the specimen surface and specularly reflected off the surface, and the wave that propagates as the surface wave along the liquid–solid interface. The two waves interfere with each other and the pattern of the interference is detected by the acoustic transducer when the two waves go back to it.

Now consider that the specimen surface initially placed at the focal point of the acoustic beam is being displaced toward the lens. As the specimen gets closer to the lens, the specular reflection path #1 is shortened by 2OC. Here O is the focal point and C is the point where a vertical line originating from point O toward the lens crosses the top surface of the specimen. Path #2, on the other hand, loses the distance by AO + OB and gains by AB. As the distance between the specimen and the lens is varied, the two acoustic waves experience constructive and destructive interferences. Accordingly, the acoustic transducer's output-voltage experiences maxima (corresponding to the constructive interference) and minima (corresponding to the destructive interference).

The voltage fluctuation associated with the interference and captured by the acoustic transducer is referred to as the V(z) curve. Figure 3.24 shows a typical V(z) curve. Since the acoustic frequency is fixed (determined by the acoustic source), the acoustic path length (the path length in the unit of the wavelength) over AB depends

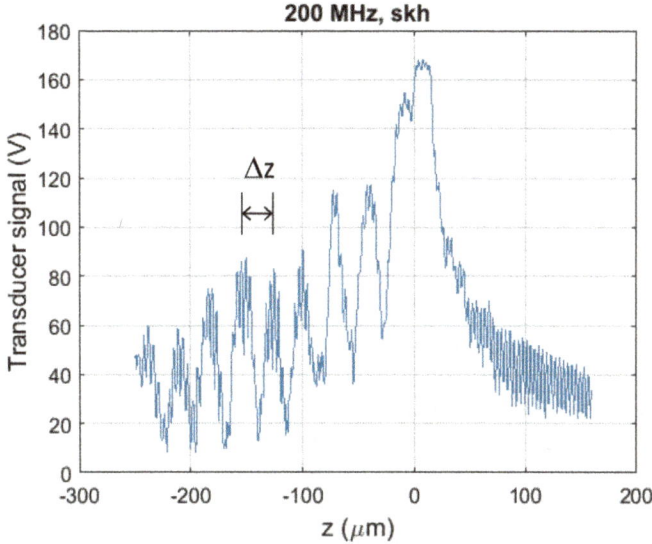

Figure 3.24. Sample V(z) curve.

on the phase velocity of the surface wave. The interval of these peaks (Δz) is associated with the velocity of the surface acoustic wave relative to the acoustic velocity in the coupling water as follows

$$V_s = \frac{V_w}{\sqrt{1 - \left(1 - \frac{V_w}{2\Delta z} \cdot f\right)^2}} \qquad (3.172)$$

where V_s is the surface acoustic wave velocity, V_w is the acoustic velocity in water and f is the acoustic frequency.

From the acoustic wave velocity V_s evaluated from the V(z) curve and equation (3.172), we can estimate the elastic constant using the fact that the acoustic velocity is proportional to the square root of the ratio of the elastic constant to the density (equation (2.13)).

3.4 Diffraction

From our daily experience, we know that we can hear sound and see light even if we are behind an obstacle. This is because as a wave, sound and light can bend over the edge of the obstacle. Particles do not have such a property; they would be blocked by the obstacle and you would not perceive them.

We can understand diffraction as a combined effect of Huygens' principle and interference. Below we discuss this effect for some instances. Although we use light waves in the examples below, the discussions here are in general applicable to other waves such as sound or water waves.

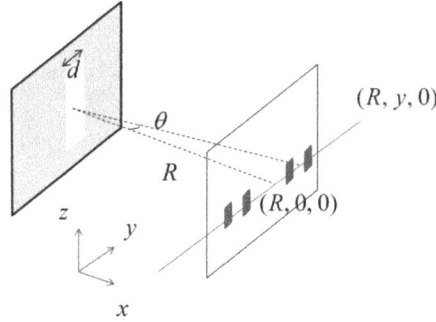

Figure 3.25. Diffraction pattern by a single slit.

Diffraction by a single slit.   First we discuss diffraction due to a single slit. Here we consider the phenomenon rather qualitatively. Those who are interested in more detailed argument are encouraged to see references such as [18].

Consider a wave of wavelength λ passing through a slit of width d in figure 3.25. Here the center of the slit is at the origin $(x, y, z) = (0, 0, 0)$, and a screen is placed normal to the x-axis at $x = R$. A diffraction pattern is formed on a screen along a line $(R, y, 0)$. θ is the angle formed by the x-axis and the line from the origin to a given point of the diffraction pattern on $(R, y, 0)$. According to Huygens' principle [19], 'every point on a wavefront is a source of wave'. In this case, the diffraction pattern on the screen results from superposition of light emitted by an infinite number of infinitesimally narrow line sources parallel to the z-axis at the slit. It is an example of Fraunhofer diffraction discussed at the end of the last chapter.

It is well known that the intensity profile of the diffraction pattern on the screen takes the following form [18]

$$I(\theta) = I(0)\left(\frac{\sin \beta}{\beta}\right)^2. \tag{3.173}$$

Here β is a function of the ratio of the slit width to the wavelength and angle θ

$$\beta = \frac{kd}{2}\sin\theta = \frac{\pi d}{\lambda}\sin\theta. \tag{3.174}$$

Figure 3.26 plots the intensity profile (3.173) as a function of $\beta/\pi = (d/\lambda)\sin\theta$ for wavelength $\lambda = 633$ nm (He–Ne laser's wavelength) with various slit widths. Notice that the intensity profile is solely determined by β/π regardless of the slit width.

Equation (3.174) indicates that for a given wavelength and slit width, β depends only on angle θ. This means that at any point on the diffraction pattern where the corresponding angle θ makes $\sin\beta/\beta = 0$, the intensity is zero. It also means that the minimum value θ_{min} that makes $\sin\beta/\beta = 0$ defines an edge of the central lobe of the intensity profile. From equation (3.173) we find that the first zero occurs when $\beta = \pm\pi$. From equation (3.174), we find that θ_{min} satisfies the following condition

$$\frac{d}{\lambda}\sin\theta_{min} = \pm 1. \tag{3.175}$$

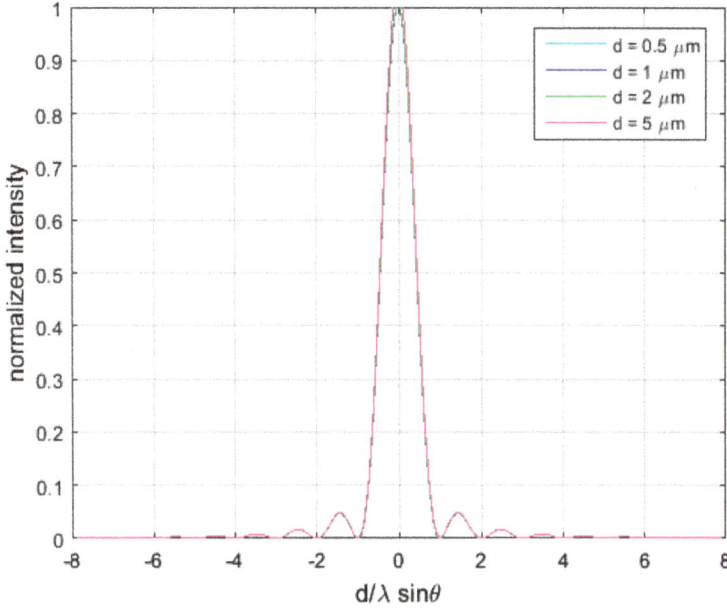

Figure 3.26. Intensity profile of diffraction by a single slit.

By solving equation, (3.175) for $\sin \theta$ and using the small angle approximation $\sin \theta \cong \theta$, we find the full angle that corresponds to the central lobe of the intensity profile as follows

$$\Delta \theta = 2\theta_{min} \cong 2 \sin \theta = 2\frac{\lambda}{d}. \qquad (3.176)$$

We can visualize the meaning of equation (3.175) as the condition for the first zero by a simple geometrical analysis. Figure 3.27 illustrates the formation of the first zero by a pair of light rays coming from the top edge and the center of the slit. We find that at angle $\theta_1 = \theta_{min}$ these two rays interfere destructively with each other on the screen. The destructive interference comes from the following equation indicating that the optical path difference between these two light rays is one half of the wavelength

$$\frac{d}{2} \sin \theta_1 = \frac{\lambda}{2}. \qquad (3.177)$$

Equation (3.177) is identical to equation (3.176). It is possible to find an infinite number of similar pairs that satisfy the condition of destructive interference as follows

$$\frac{d}{2} \sin \theta_m = m\frac{\lambda}{2}, \qquad i.e. \qquad \frac{d}{2m} \sin \theta_m = \frac{\lambda}{2}. \qquad (3.178)$$

Here m is an integer. The first part of equation (3.178) explicitly indicates that this condition leads to $d \sin \theta = m\lambda$, which in turn makes $\beta = m\pi$ via equation (3.174),

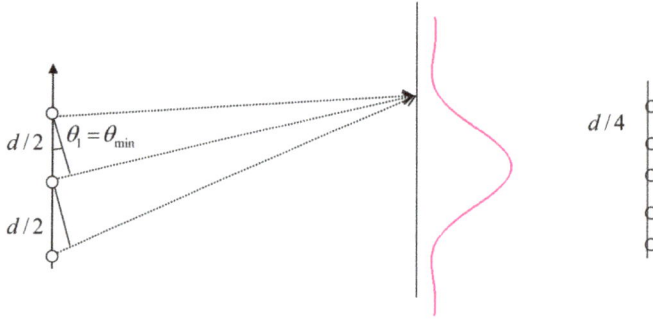

Figure 3.27. Formation of first zero by a pair of light rays.

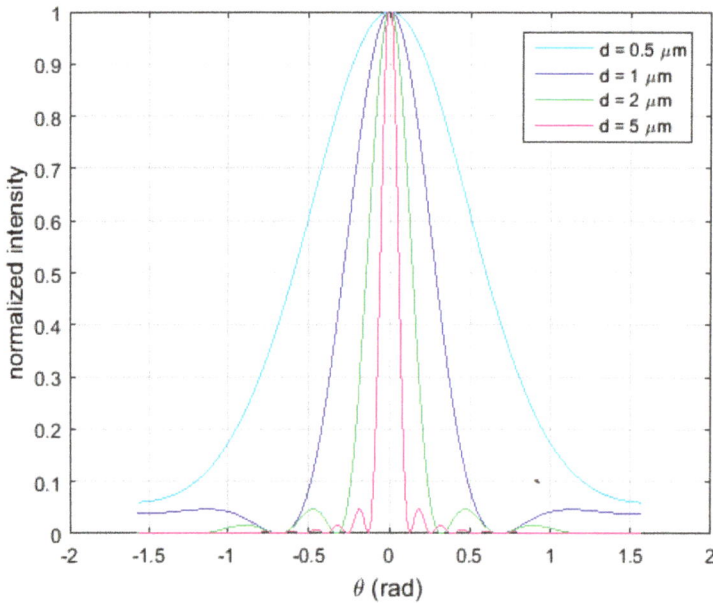

Figure 3.28. Intensity profile of diffraction by a single slit as a function of θ.

hence $\sin \beta = 0$ in equation (3.173). The second part of equation (3.178) indicates that mth zero is formed by another pair of line sources which are closer to each other by a factor of m as compared with the pair of sources that form the first zero. The inset in figure 3.27 illustrates the $m = 2$ case due to line sources separated by $d/4$.

Figure 3.28 plots the intensity as a function of θ for the same slit widths and wavelength as figure 3.26. It clearly illustrates the fact that the central lobe of the intensity profile increases as the slit width decreases. Note that the broadest intensity profile in figure 3.28 is the case where the wavelength (633 nm) is longer than the slit width (500 nm). The other profiles are when the wavelength is shorter than the slit width.

Diffraction by a circular aperture and diffraction limit. Similar to the single slit case, the intensity of diffraction due to a circular aperture is given as follows

$$I(\theta) = I(0)\left(2\frac{j_1(\beta)}{\beta}\right)^2. \tag{3.179}$$

Here $j_1(\beta)$ is the first order Bessel function of first kind and β is as defined by equation (3.174). The first zero occurs when $\beta = 3.8317$. It follows that the following condition defines the central lobe in this case

$$\Delta\theta \approx \sin\theta_{min} = 3.8317\frac{\lambda}{\pi d} = 1.22\frac{\lambda}{d}. \tag{3.180}$$

Equation (3.180) corresponds to equation (3.176). Similar to the single slit case, the size of the central lobe of the diffraction pattern is determined by the wavelength relative to the slit size, and the central lobe is defined by the first zero associated with the minimum angle θ_{min}. From this minimum angle, we can derive an important concept known as diffraction limit.

Figure 3.29 draws the ray from the center of aperture to the first zero. Call the radius of the central lobe (distance from the center of the central lobe to the first zero) a and the distance from the aperture to the screen R. From figure 3.29, we find that $a = R\tan\theta_{min} \cong R\sin\theta_{min}$. Thus, using equation (3.180), we can express the radius of the central lobe as follows

$$a = 1.22\frac{R\lambda}{d}. \tag{3.181}$$

Equation (3.181) tells us that light of wavelength λ going through an aperture with diameter d has a minimum radius a at the screen R (m) away. In the context of focusing a laser beam with a positive lens of diameter D and focal length f, we can use $R = f$, $d = D$. Thus, we obtain the following equation for the minimum diameter $2a$ of the laser beam at the focus

$$2a = 2 \times 1.22\frac{f\lambda}{D} = 2.44F\lambda. \tag{3.182}$$

Here $F = f/D$ is called the F number of the lens.

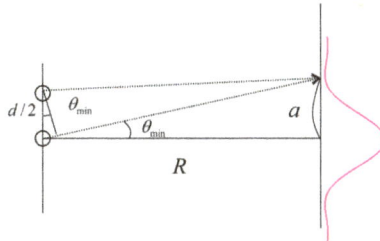

Figure 3.29. Diffraction limit.

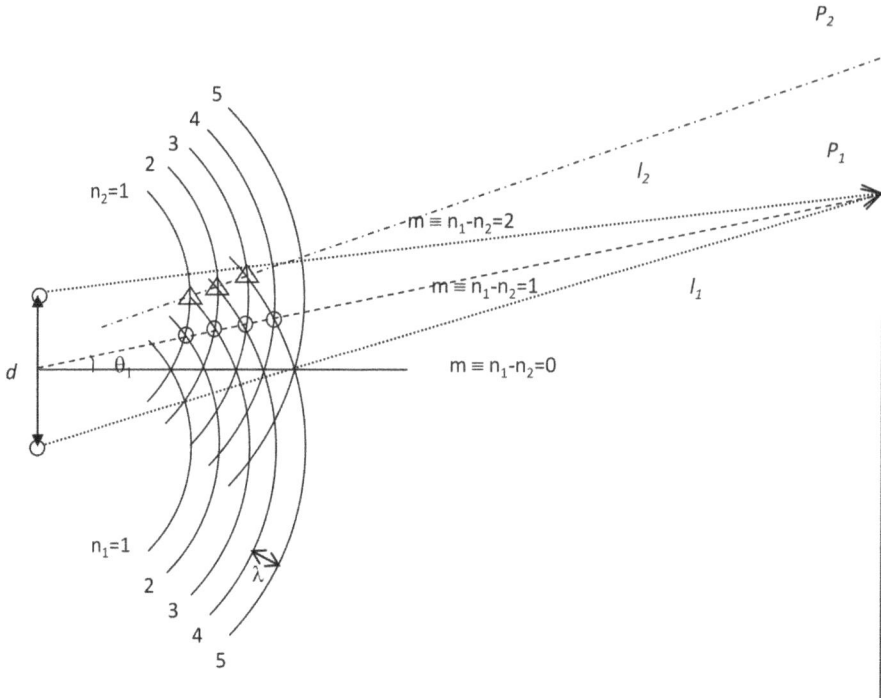

Figure 3.30. Transmissive diffraction grating.

Diffraction grating. A diffraction grating consists of a number of equally spaced grids that block light. There are reflective and transmissive types but the principle of operation is the same. We use a transmissive type here with figure 3.30. In this figure, the light is incident from the left and blocked by the grids spaced by d. Each space between the neighboring pair of grids can be viewed as a slit. The series of curved wavefronts originating from each slit interfere with the series of wavefronts originating from other slits. We assign an integer to the wavefront (say the crest) from the slit side as 1, 2, ... n. Each wavefront interferes with other wavefronts, and the number of combinations is infinite. When the optical path length difference between a given pair of wavefronts is an integer multiple of the wavelength, the interference is constructive.

Consider such a constructive interference using the wavefronts from the central slit. The first wavefront from the central slit can interfere with the first wavefront one slit away (the next slit). Similarly, the second (nth) wavefront from the central slit can interfere with the second (nth) wavefront from the next slit. In these cases, the difference in the wavefront number (the integer assigned above) is zero, and referred to as zeroth order diffraction. If we connect all these points where the wavefronts from the central slit form zeroth order diffraction, we can draw a straight line perpendicular to the plane of the diffraction grading.

In the same fashion, we can consider first-order diffraction; the interference is based on the nth wavefront from the central slit with ($n + 1$) wavefront from the next

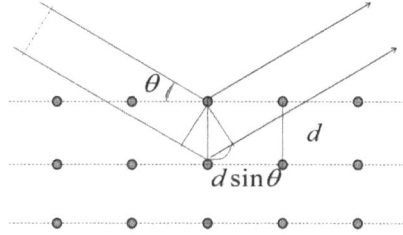

Figure 3.31. X-ray diffraction.

slit. By connecting all these points, we can draw a straight line going away from the grating surface at a certain angle. Call this angle θ_1 as it represents first order diffraction. We can draw straight lines of diffractions of other orders; θ_m for the mth order diffraction based on n and $(n + m)$ wavefronts.

Referring to figure 3.30, we can express the condition of the constructive interference as follows

$$d \sin \theta_m = m\lambda. \tag{3.183}$$

Equation (3.183) is known as the formula for the diffraction angle. Normally, the grid spacing d is known. By measuring the angle of diffraction(s), we can estimate the wavelength of the light incident to the diffraction grating. This technique is used to identify the source of light.

X-ray diffraction. X-ray diffractometry is an application of diffraction. Crystals are arrays of parallel atomic planes, as illustrated by figure 3.31. When an x-ray is incident to the surface of a crystal, it is scattered by atoms forming a diffraction pattern in the same fashion as a reflective diffraction grating. The condition to form constructive interference, known as Bragg's law, is given as follows

$$d \sin \theta = m\lambda. \tag{3.184}$$

Here d is the interatomic-plane distance and θ is the angle made by the crystal surface and the incident x-ray. By using an x-ray of a known wavelength and measuring the angle θ, we can estimate the interatomic-plane distance from equation (3.184).

One application of this technique is the estimation of residual stresses. If a residual stress caused by a certain process such as cutting or welding causes the interatomic-plane distance to become shorter or longer, we can detect it.

3.5 Doppler effect

The Doppler effect is observed in any type of wave when the source moves toward or away from the detector. When you hear a police siren while you are driving, you feel uneasy if the pitch of the siren goes up. From experience, you know that it indicates that the police car is getting closer to you. On the contrary, if the pitch of the siren goes down, you know that the distance between your car and police car is increasing. This effect is known as the Doppler effect; when the source is approaching the sensor the frequency sensed by the sensor goes up and when the source is going away from the sensor the frequency sensed by the sensor goes down.

The Doppler effect is not caused by an actual change in the frequency of the source. You can easily imagine that the police officer would hear the siren at the same pitch all the time. The reason for the change in the frequency detected by the sensor is that the total number of waves that the sensor captures increases if the source is approaching, and that the total number of waves captured by the sensor decreases if the source is going away from the sensor. By definition, the number of waves sensed in a unit time is the frequency. So, the velocity of the sensor relative to the source changes the frequency of wave detected by the sensor. This change in frequency is called the Doppler shift. Generally, we can express Doppler shifted frequency as follows

$$f = \left(1 + \frac{\Delta v}{c}\right) f_0 \qquad (3.185)$$

$$\Delta f = \frac{\Delta v}{c} f_0. \qquad (3.186)$$

Here f is the frequency with Doppler shift of Δf, Δv is the relative velocity between the source and sensor and c and f_0 are the velocity and frequency of the incoming wave. When the relative velocity is much smaller than the incoming wave velocity ($\Delta v \ll c$), the Doppler shift is negligible.

Doppler effect is important in scientific phenomena as well. Here I would like to discuss two topics as examples. The first is known as Doppler broadening of spectral width often observed in gas lasers. In section 3.2.3, we discussed the electric susceptibility. Normally, as we observed in that section, electrons respond to the electric field of incoming light most efficiently at its natural frequency. Equation (3.126) expresses the response quantitatively. Also, figure 3.16 illustrates the response graphically where the peak frequency is the natural frequency. When the pressure of the gain medium in a gas laser is low, the active molecules responsible for the laser gain move at high speed comparable to the speed of light. This causes each dipole pair to feel the Doppler shift according to equation (3.185), meaning that some dipole pairs respond at $f = f_0 - \Delta f$ while other pairs respond at $f = f_0 + \Delta f$. The differential velocity Δv has some statistical broadening obeying the Maxwell distribution at the gas temperature. Consequently, the peak frequency of the response function like figure 3.16 becomes broadened. This broadening is referred to as Doppler broadening. A typical He–Ne laser oscillating at 632.8 nm has Doppler broadening of the order of GHz [20].

The other topic I would like to discuss is the technique called Doppler cooling [21, 22]. The idea is as follows. You apply a light beam to atoms from opposite sides. The frequency of the light is slightly lower than the absorption frequency of the atom. Consequently, those atoms moving toward the photon can only absorb the photon (because due to the Doppler shift the optical frequency is tuned to the absorption for these atoms). By the absorption, the photon loses the momentum. From the momentum conservation, the momentum of the atom after the absorption must be lower in the direction of the original velocity of the atom (because the atom was moving in the opposite direction to the photon). In this fashion, eventually the atom can lose its velocity, or cool down.

References

[1] Hecht E 2002 *Optics* 4th edn (San Francisco, CA: Addison Wesley) chapter 4, sections 4.4 and 4.6

[2] Kinsler L E, Frey A R, Coppens A B and Sanders J V 1980 *Fundamentals of Acoustics* 3rd edn (New York: Wiley) ch 6, sec 6.4

[3] Graff K F 1975 *Wave Motion in Elastic Solids* (New York: Dover) ch 6, sec 6.1

[4] Kinsler L E, Frey A R, Coppens A B and Sanders J V 1980 *Fundamentals of Acoustics* 3rd edn (New York: Wiley)

[5] Graff K F 1975 *Wave Motion in Elastic Solids* (New York: Dover)

[6] Kinsler L E, Frey A R, Coppens A B and Sanders J V 1980 *Fundamentals of Acoustics* 3rd edn (New York: Wiley) ch 10, sec 10.4

[7] Griffiths D J 1999 *1999 Introduction to Electrodynamics* 3rd edn (Upper Saddle River, NJ: Prentice Hall) ch 5 and 7

[8] Boas M L 2006 *Mathematical Methods in the Physical Sciences* 3rd edn (London: Wiley) ch 6

[9] Slater J C and Frank N H 1947 *Electromagnetism* (New York: McGraw-Hill)

[10] Rose J L 1999 *Ultrasonic Waves in Solid Media* (Cambridge: Cambridge University Press) ch 2

[11] Griffiths D J 1999 *Introduction to Electrodynamics* 3rd edn (Upper Saddle River, NJ: Prentice Hall) 1999 ch 9

[12] Crandall I B 1926 *Theory of Vibrating Systems and Sound* (New York: van Nostrand)

[13] Slater J C and Frank N H 1947 *Electromagnetism* (New York: McGraw-Hill) ch 11

[14] Misner C W, Thorne K S and Wheeler J A 1973 *Gravitation* (New York: W H Freeman and Company)

[15] Barish B C and Weiss R 1999 LIGO and detection of Gravitational waves *Phys. Today* **52** 44–50

[16] Einstein A 1954 *The meaning of relativity* 5th edn (New York: M JF Books)

[17] Einstein A 1911 On the influence of gravitation on the propagation of light *Annalen der Physik* **35**

[18] Hecht E 2002 *Optics* 4th edn (San Francisco: Addison Wesley) ch 10

[19] Hecht E 2002 *Optics* 4th edn (San Francisco: Addison Wesley) ch 4

[20] Yariv A 1971 *Introduction to Optical Electronics* (New York: Holt, Rinehart and Winston) ch 5

[21] Wineland D J and Dehmelt H 1975 Proposed $10^{14}\Delta v < v$ laser fluorescence spectroscopy on Tl$^+$ mono-ion oscillator III *Bull. Am. Phys. Soc.* **20** 637

[22] Hansch T W and Shawlow A L 1975 Cooling of gases by laser radiation *Opt. Commun.* **13** 68

Chapter 4

Wave propagation

4.1 Group velocity

In section 3.2, we observed that in certain physical systems the phase velocity of a wave depends on the frequency. This type of physical system is generally referred to as a dispersive medium. When a wave travels through a nondispersive medium, the phase velocity is constant regardless of the frequency. Therefore, even if multiple waves of different frequencies travel simultaneously, they all travel at the same speed. However, if the same multiple waves travel in a dispersive medium, each wave travels at its own velocity. In this situation, we cannot define a phase velocity. Instead, we use group velocity.

Consider the relation between the angular frequency ω and wavenumber k for a dispersive medium in figure 4.1 where the former is expressed as a function of the latter, $\omega(k)$. The solid curved line represents $\omega(k)$. For a given k, the phase velocity $v_p(k)$ and group velocity $v_g(k)$ are defined as follows

$$v_p = \frac{\omega}{k} \qquad (4.1)$$

$$v_g = \left(\frac{d\omega}{dk}\right)_\omega. \qquad (4.2)$$

Figure 4.1 illustrates that at two representative points $(\omega, k) = (\omega_1, k_1)$ and $(\omega, k) = (\omega_2, k_2)$, we can find the phase velocity and group velocity as the slope of the line connecting the origin and the point, and as the slope of the $\omega(k)$ curve at that point.

Consider some behaviors of group velocity in association with dispersion. Rewrite equation (4.1) in the following form

$$\omega = kv_p. \qquad (4.3)$$

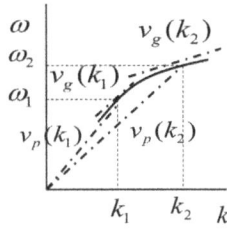

Figure 4.1. Phase velocity and group velocity as a function of wavenumber.

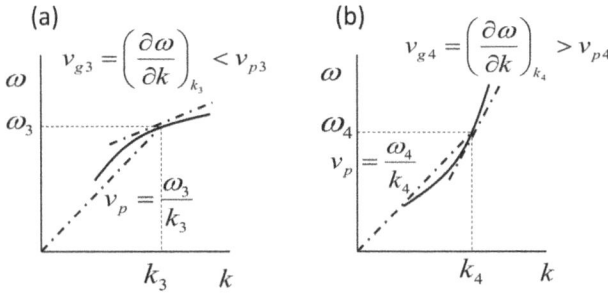

Figure 4.2. (a) Normal and (b) anomalous dispersion.

Substituting equation (4.3) into equation (4.2), we find

$$v_g = \frac{d(k v_p)}{dk} = v_p + k \frac{dv_p}{dk}. \tag{4.4}$$

Equation (4.4) indicates that if $dv_p/dk < 0$, $v_g < v_p$ and if $dv_p/dk > 0$, $v_g > v_p$. The former case is referred to as normal dispersion and the latter anomalous dispersion. Figure 4.2 illustrates (a) normal dispersion and (b) anomalous dispersion at representative points $(\omega, k) = (\omega_3, k_3)$ and $(\omega, k) = (\omega_4, k_4)$, respectively. Notice that if $\partial^2\omega/\partial k^2 < 0$ the dispersion is normal and if $\partial^2\omega/\partial k^2 > 0$ the dispersion is anomalous.

In the case of light, the dispersion comes from the fact that the index of refraction depends on the wavenumber. Expressing the index of refraction as a function of wave number $n(k)$ and using $v_p = c_0/n$ (c_0: speed of light in vacuum), we can rewrite equation (4.4) as follows

$$v_g = \frac{c_0}{n} + k \frac{d}{dk} \frac{c_0}{n(k)} = \frac{c_0}{n} - \frac{c_0 k}{n^2} \frac{dn}{dk}. \tag{4.5}$$

Equation (4.5) explicitly indicates that the group velocity depends on the wavenumber dependence of the index of refraction. If $dn/dk > 0$ ($dn/d\omega > 0$), $v_g < c$ or the dispersion is normal. If $dn/dk < 0$ ($dn/d\omega < 0$), $v_g > c$ or the dispersion is anomalous. Figure 4.3 indicates the frequency range where the dispersion is anomalous.

Simple demonstration of group velocity. In the above discussion, the concept of group velocity was somewhat abstract. Here we visualize the concept using a simple

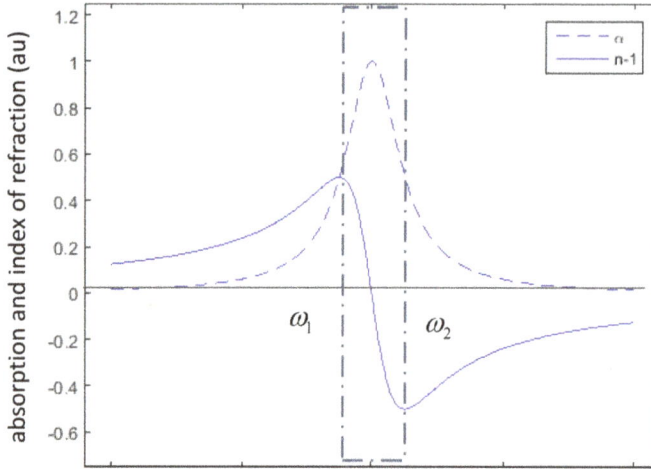

Figure 4.3. Index of refraction as a function of angular frequency. Box indicates frequency range where dispersion is anomalous.

model. Group velocity becomes significant when a group of waves with more than one frequency propagate in a dispersive medium. If the wave has a single frequency, even if the phase velocity is a steep function of frequency in the dispersive medium, the wave propagates at a certain phase velocity. In reality, the frequency of the wave from a source has a certain range of frequency. This range is called the band-width as in the frequency domain it corresponds to a certain width of frequency. For example, if you use an output beam from a laser source, its frequency has a certain width in Hz. As an example, He–Ne laser's band width is typically 1 GHz [1]. If the laser beam propagates in a dispersive medium whose dispersion property is such that the velocity at the red end (the lowest frequency end) is considerably different from the blue end (the highest frequency end), the group velocity becomes meaningful.

We can consider a simple case using two waves having slightly different frequencies. Recall the beat phenomenon we discussed in chapter 2. There we did not consider any difference in the phase velocity at ω_1 and ω_2. However, if the two waves are traveling in a dispersive medium and the phase velocity is different at ω_1 and ω_2, we can argue the group velocity as the velocity of the envelope generated by the beat.

Figures 4.4–4.6 illustrate sample cases where two waves of similar frequency travel in three types of media. The three media exhibit nondispersive, normal dispersive, and anomalous dispersive characteristics, respectively. The normal and anomalous dispersive characteristics are hypothetical. In all three figures the first wave has 10% lower frequency than the second wave, and the three graphs are snapshots of the combined wave at three different times. Since the frequencies of the two waves are close to each other, we see envelopes due to the beats in all three figures.

In figure 4.4, the two waves travel in a nondispersive medium. The two waves have the same phase velocity 0.2 m s^{-1}. (I use the unit of m for the length and s for

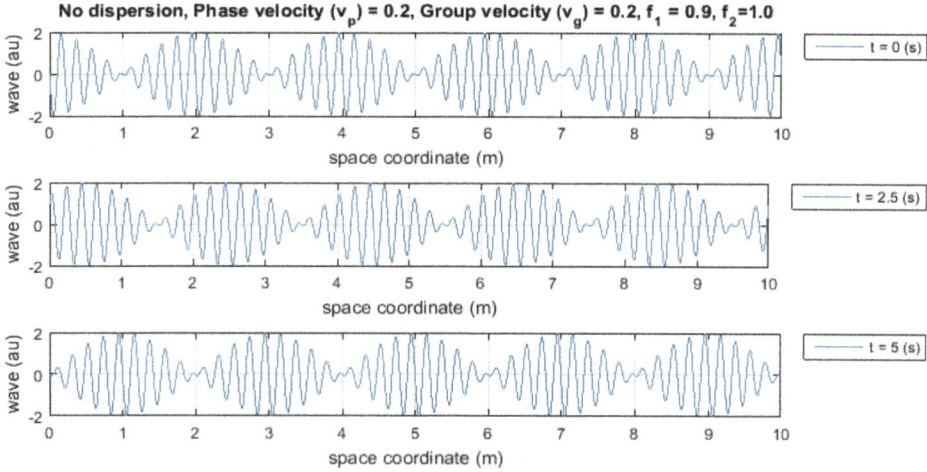

Figure 4.4. Two sinusoidal wave having similar frequencies with no dispersion. Units: length in m, time in s, velocity in m s^{-1}.

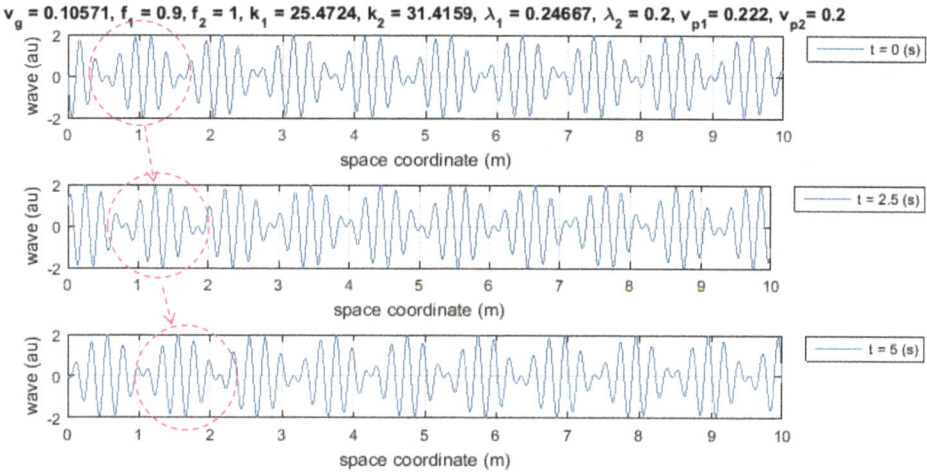

Figure 4.5. Two sinusoidal waves having similar frequencies with positive group velocity. Units: length in m, time in s, velocity in m s^{-1}.

the time but the choice of the units is arbitrary. There is no physical meaning.) The beat envelope also travels at velocity 0.2 m s^{-1}. We can say that in this case the group velocity is the same as the phase velocity because the medium is dispersionless. In figure 4.1, we can easily see that if $\omega(k) = v_p k$, the local slope at any point is the same as the slope of the function itself.

In figure 4.5, the two waves experience normal dispersion. The frequency and wavenumber of the first and second waves are, respectively, $f_1 = 0.9$ Hz, $k_1 = 25.47$ m^{-1}, and $f_2 = 1.0$ Hz, $k_2 = 31.42$ m^{-1}. From equation (4.1), the phase velocities are $v_{p1} = 0.22$ m s^{-1} and $v_{p2} = 0.2$ m s^{-1}. From equation (4.2), we can find the group

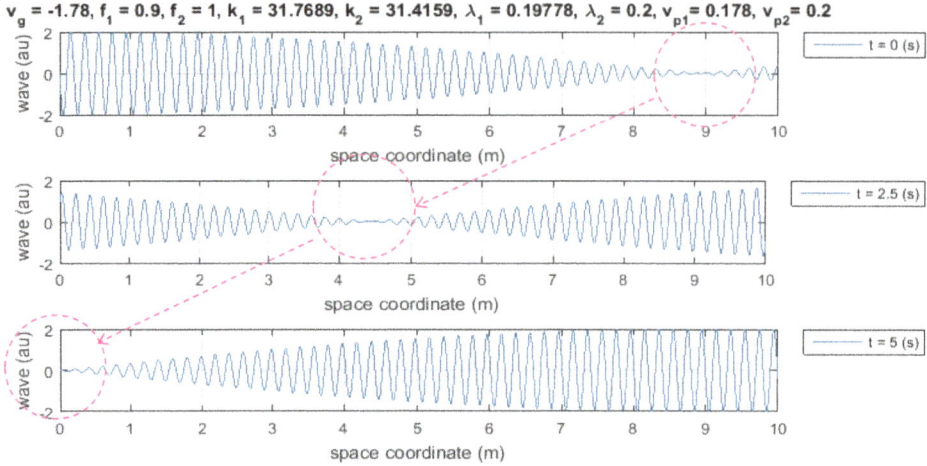

Figure 4.6. Two sinusoidal waves having similar frequencies with negative group velocity. Units: length in m, time in s, velocity in m s^{-1}.

velocity as $v_g = \Delta\omega/\Delta k = 2\pi \times (1 - 0.9)/(31.42 - 25.47) = 0.1057$ m s^{-1}. Thus, $v_g < v_{p2} < v_{p1}$, hence the dispersion is normal. The three graphs clearly demonstrate this group velocity; in 5 s, the circled envelope travels approximately 0.5 m ($\cong 0.5/5 = 0.1$ m s^{-1}).

The two waves in figure 4.6 experience negative dispersion. Repeating the same calculation as for figure 4.5, we find the phase and group velocities as $v_{p1} = 0.178$ m s^{-1}, $v_{p2} = 0.2$ m s^{-1} and $v_g = -1.78$ m s^{-1}. The circled portion of the beat wave clearly shows this negative group velocity.

4.2 Resonator

In chapter 1, we briefly discussed a standing wave on a string. The standing wave was formed when two counter-traveling waves of the same amplitude and phase velocity were superposed with each other. The necessary conditions in this case were that at the far end of the string, the wave was reflected with zero amplitude (fixed end reflection) (equation (1.85)) and the length of the string was an integer multiple of half wavelength (equation (1.86)). We can view this situation as the wave energy being stored in the string under resonance.

The system that stores wave energy in this fashion is referred to as a resonator, because the wave dynamics resonates with the property of the system. In the case of the standing wave on a string, equation (1.86) describes the resonant condition in terms of the relation between the wavelength and the string length. The frequency is determined in accordance with this wavelength and the phase velocity determined by the tension of the string. Under these conditions, the two ends of the string keep reflecting the wave towards the other end.

Similar phenomena are observed in other waves in general. Figure 4.7 depicts the situation schematically. An incident wave approaches the resonator from the left. Part of the wave passes through the left end of the resonator towards the right end.

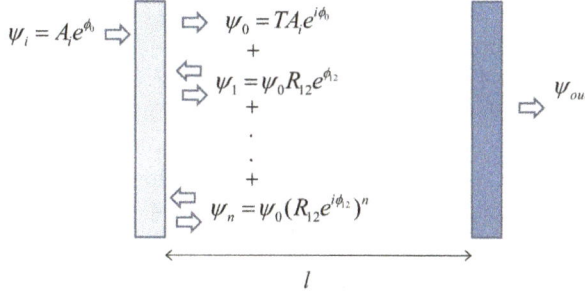

Figure 4.7. Waves in a stable resonator.

This rightward wave is reflected off the right end of the resonator and travels leftward to the left end of the resonator. There, the wave is reflected again towards the right end. If the incident wave stops coming to the resonator, this back-and-forth motion of the wave eventually dies out due to the loss mechanism (such as air resistance or friction in the case of the string wave). However, if the incident wave provides the resonator with energy just enough to compensate the loss, the dynamics continues steadily until the incident wave stops.

Referring to figure 4.7, we can express the intra-resonator wave for each round trip as follows

$$\psi_0 = T_1 A_i e^{i\phi_0} \tag{4.6}$$

$$\psi_1 = \psi_0 R_{12} e^{i\phi_{2l}} \tag{4.7}$$

$$\psi_2 = \psi_1 R_{12} e^{i\phi_{2l}} = \psi_0 (R_{12} e^{i\phi_{2l}})^2$$
$$\cdots \tag{4.8}$$
$$\psi_n = \psi_n (R_{12} e^{i\phi_{2l}})^n.$$

Here T_1 is the transmission of the first reflector, R_{12} is the product of the reflection of the two reflectors and ϕ_{2l} is the phase change for each round trip for the resonator length l. Equation (4.6) is the wave about to leave the first reflector after the incident beam enters the resonator for the first time. Equation (4.7) is the wave leaving the left reflector after the first round trip. Similarly, equation (4.8) is the wave about to leave the left reflector after n-round trips inside the resonator. We can see that after each round trip, factors of R_{12} and $exp(i\phi_{2l})$ are multiplied to the amplitude and phase, respectively. Combining these factors and taking the sum of all waves at the inside of the left reflector, we can find the following expression for the total wave leaving the left reflector towards the right reflector the resonator

$$\psi_{tot} = \psi_0 \frac{1 - (R_{12} e^{i\phi_{2l}})^n}{1 - (R_{12} e^{i\phi_{2l}})} \approx \psi_0 \frac{1}{1 - (R_{12} e^{i\phi_{2l}})} = \frac{T_1 A_i}{1 - (R_{12} e^{-i2kl})}. \tag{4.9}$$

In the right-hand term of equation (4.9), the initial phase $\phi_0 = 0$ is set to 0 ($e^{i\phi_0} = 1$) and the round trip phase is replaced with the propagation constant k and the resonator length l with the time term omitted in $i\phi = i(\omega t - kl)$).

If the resonator length is such that the wave leaving the left reflector after i round trips has the same phase as the wave after $i - 1$ round trips (i.e. the wave after a given round trip has the same phase as the ones after previous round trip), all the waves leaving the left reflector are in-phase. Under this condition, it is possible to establish the equilibrium situation where the newly incoming power (provided by the source or the incident wave) cancels the total power loss in the resonator. In other words, the newly incoming power compensates all the losses, including the power going out from the right reflector (see below). If such an equilibrium condition is established, we can store the wave energy stably in the resonator. We call this type of resonator a stable resonator and say that the incident wave is coupled with the resonator.

Note that if the right reflector has some transmission, the wave passes through the resonator. The energy going out of the resonator through the right reflector can be viewed as part of the intra-resonator loss and compensated by the source. Here the important thing is that only the waves that satisfy the above phase and wavelength conditions can pass through the resonator. If the incident wave has components of other wavelengths or phase conditions, those components do not survive the resonator. From this standpoint, the resonator acts as a filter. Since this type of resonator does not generate energy, it is called a passive resonator. It is also possible to prove that under this resonant condition, the phase condition at the left reflector is such that the reflection of the incident wave is canceled out by the intra-resonator wave.

When the right reflector has transmission, the output wave is as follows

$$\psi_{\text{out}} = \frac{T_{12} A_i}{1 - (R_{12} e^{-i2kl})}. \tag{4.10}$$

Here T_{12} is the product of the transmission of the two reflectors. The phase corresponding to passing the resonator once from the left to right is omitted in equation (4.10) as it is common with all the waves leaving the left mirror.

If the resonator has a mechanism to generate wave energy, it can replace the source energy. This type of resonator is called an active resonator. A laser oscillator is an example of active resonators. Interested readers are encouraged to read references [1, 2]. These references describe the above-mentioned phase conditions and other conditions related to resonance for optical resonators. Reference [3] explains an optical resonator as an extremely narrow band filter.

FFT method A technique known as the FFT (fast Fourier transform) method [4] is useful to analyze the wavefront profile numerically. Reference [4] discusses application of the FFT method to analysis of the laser interferometric gravitational wave detector discussed in section 3.3.1. Miyasaka *et al* [5] apply the FFT technique to analyze the ultrasound beam propagating in a scanning acoustic microscope discussed in section 3.3.2.

The FFT method evaluates the transverse profile of a wave in the spatial frequency domain as the wave propagates. To find the transverse profile, we first compute the Fourier transform of the amplitude at a reference point of the propagation using the fast Fourier transform (FFT) algorithm. We then multiply the Fourier transform to a quantity known as the propagator to evaluate the Fourier transform of the transverse profile at a given point of the propagation (an intermediate point). Any factor that causes a change in the wavefront profile is taken into account in the frequency domain by multiplying the corresponding transfer function to the Fourier transform of the wavefront. For instance, if a wave is passing through an aperture between the initial and intermediate point, we can multiply the frequency domain expression (transfer function) of the aperture along with the propagator to reach that point. We can repeat this procedure for all the factors that affect the wavefront profile at each intermediate point. At the final point, we can inversely fast Fourier transform the Fourier transform of the wavefront at that point. The resultant space-domain data represents the transverse profile of the wave at the destination point. In this fashion, we can consider any distortion on the transverse profile, e.g. part of the wave is blocked by a small obstacle. The FFT method is generally faster than the corresponding modal analysis.

Consider a paraxial light beam traveling in the positive z-direction. The FFT method [4] expands the wavefront at a given point of propagation on the z-axis into a series of plane waves traveling in different directions. To consider the transverse profile along the x-axis, the direction of each plane wave can be represented by the angle from the z-axis, θ_i. Here the index i denotes the ith direction. Mathematically, this is equivalent to expanding the beam profile to a Fourier series whose basis is the x-component of the propagation vector k, $k_x = k \sin \theta_i$ (figure 4.8 (a)). As the beam propagates in the z-direction, the ith plane wave propagates in the direction of θ_i, carrying the amplitude represented by the ith component of the Fourier series. Each component has different propagation characteristics from one another. At another point on the z-axis, we can find the transverse profile by inversely Fourier transforming the Fourier series at that point resulting from the propagation of all the component plane waves. The same argument holds for the beam profile along the y-axis. In the frequency domain, this propagation is expressed by propagator in the form of equation (4.11)

$$h(x, y) = e^{iz \frac{k_{x_i}^2 + k_{y_i}^2}{2k}}. \tag{4.11}$$

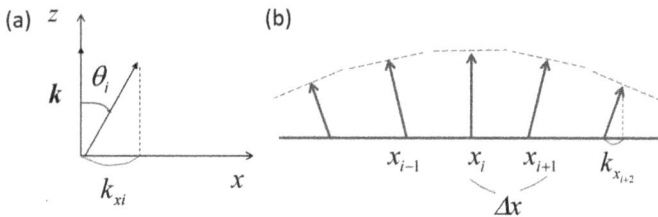

Figure 4.8. (a) Propagation vector component of ith plane wave and (b) propagation of plane wave components for a spherical wave.

Here z is the distance of propagation, kx_i and ky_i are the x and y components of the propagation vector for the ith plane wave.

Now consider the application of the FFT method to a Gaussian beam (including higher order Gaussian beams). For simplicity, here we consider the problem as a one-dimensional case, but the same argument holds for two-dimensional cases. As a spherical wave, going from the beam center toward the edge, the direction of the propagation, which is perpendicular to the spherical wave front, deviates from the z-axis. In other words, the angle θ_i increases in proportion to the square of the radius, so does the x-component of the propagation vector (figure 4.8 (b)). This means that for a given spot size, the smaller the radius of curvature the greater the x-component of the propagation vector. Consequently, in the frequency domain, the smaller the radius of curvature the higher the maximum frequency becomes, requiring a high spatial sampling rate. For a given window size (the size of the region in the spatial domain to compute the Fourier transform), the higher grid point number is required to evaluate the Fourier transform accurately. Reference [6] describes application of the FFT method to laser interferometric gravitational wave detectors. Reference [7] discusses the effect of grid point number on application of the FFT method to steeply varying wavefronts.

4.3 Modulation

4.3.1 Amplitude modulation

Amplitude modulation is a technique commonly used in electronic communication to send a signal as a temporal variation of the amplitude of a sinusoidal wave. The frequency of the wave is called the carrier frequency (because it carries the signal) and normally the frequency of the amplitude modulation is lower than the carrier frequency. Consider a case where the amplitude is modulated sinusoidally

$$\psi_{am} = A\big(1 + \alpha \cos 2\pi f_m t\big) \cos 2\pi f_c t = A(1 + \alpha \cos \omega_m t) \cos \omega_c t. \tag{4.12}$$

α indicates the degree of the modulation referred to as the modulation depth. With the trigonometric identities, we can rewrite equation (4.12) as follows

$$\psi_{am} = A \cos \omega_c t + \frac{\alpha A}{2}(\cos(\omega_c + \omega_m)t + \cos(\omega_c - \omega_m)t). \tag{4.13}$$

Equation (4.13) explicitly indicates that the amplitude modulation generates two frequency components equally separated from the central (carrier) frequency on the high and low frequency sides.

Figure 4.9 shows sample amplitude-modulated signals in the time and frequency domains for two different modulation depths. The two signals appearing on the high and low sides of the carrier frequency are called the sidebands. The frequency domain signals indicate that the frequencies of the sidebands are equally distanced from the central frequency and their mutually equal peak hight increases with the modulation depth. This is exactly what equation (4.13) indicates. The two terms inside the parenthesis on the right-hand side of this equation represent the two

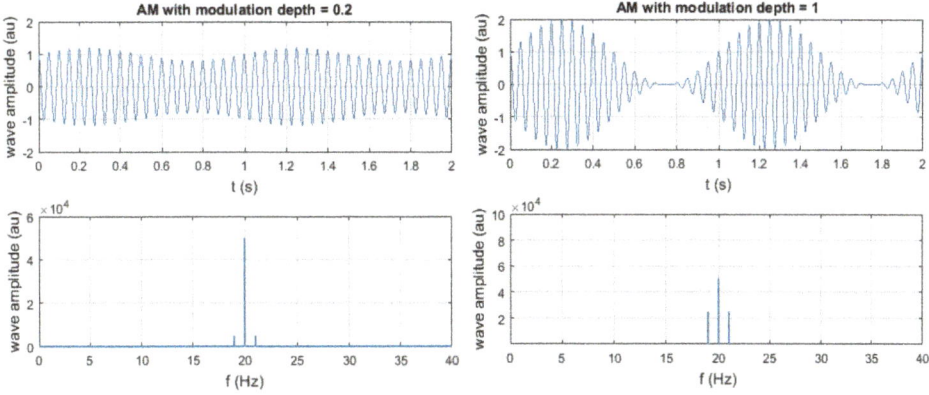

Figure 4.9. Amplitude modulation.

sidebands. A common factor of $\alpha A/2$ is multiplied to the parenthesis and the two sideband frequencies are $\pm\omega_m$ different from the carrier frequency ω_c in the arguments of the two sine terms inside the parenthesis.

Note that when the modulation depth is unity, the time domain signal indicates that at the modulation frequency the amplitude modulated signal takes zero. The reason for this is clear from equation (4.12). If $\alpha = 1$, $\cos\omega_m t$ takes the minimum value of -1 every period. At this moment, the $\alpha \cos\omega_m t$ term completely cancels 1 in the parenthesis.

4.3.2 Phase modulation

Put the original (unmodulated) wave in the following form

$$\psi = A\cos(2\pi f_c t + \phi_m(t)) = A\cos(\omega_c t + \phi_m(t)). \tag{4.14}$$

Here the frequency (called the carrier frequency) f_c is constant and ϕ_m is the phase modulated with a certain mechanism.

Put the modulated phase term as follows

$$\phi_m(t) = \gamma\sin(\omega_m t). \tag{4.15}$$

where γ indicates the degree of modulation, called the modulation index (depth).

Using equation (4.15), we can write the modulated wave in a compact form as equation (4.16)

$$\begin{aligned}\psi &= A\cos(\omega_c t + \gamma\sin(\omega_m t)) \\ &= A\cos\omega_c t\cos(\gamma\sin(\omega_m t)) - A\sin\omega_c t\sin(\gamma\sin(\omega_m t)).\end{aligned}$$

Here

$$\begin{aligned}\cos(\gamma\sin(\omega_m t)) &= J_0(\gamma) + 2J_2(\gamma)\cos(2\omega_m t) + 2J_4(\gamma)\cos(4\omega_m t) + \cdots \\ \sin(\gamma\sin(\omega_m t)) &= 2J_1(\gamma)\sin(\omega_m t) + J_3(\gamma)\sin(3\omega_m t) + \cdots.\end{aligned} \tag{4.16}$$

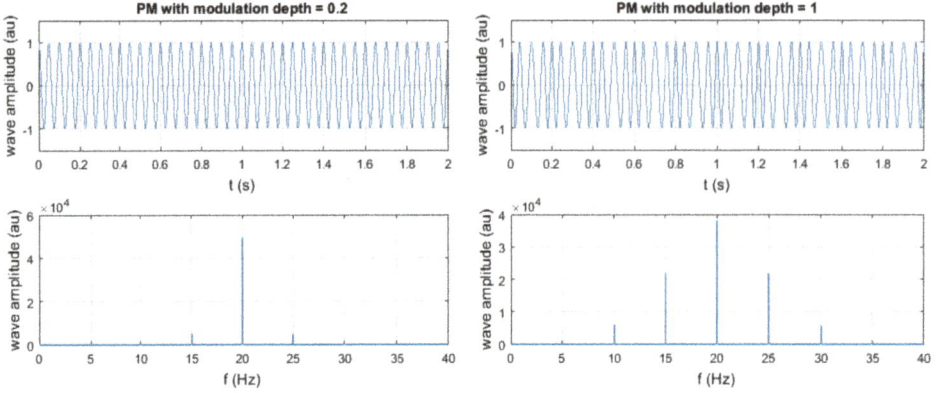

Figure 4.10. Phase modulation.

Substituting the two identities in (4.16) into equation (4.16), we obtain as follows

$$
\begin{aligned}
\psi &= A[J_0(\gamma) \cos \omega_c t + 2J_2(\gamma) \cos \omega_c t \ \cos (2\omega_m t) \\
&\quad + 2J_4(\gamma) \cos \omega_c t \ \cos (4\omega_m t) + \cdots] \\
&\quad - 2J_1(\gamma) \sin \omega_c t \sin (\omega_m t) - 2J_3(\gamma) \sin \omega_c t \sin (3\omega_m t) + \cdots] \\
&= A[J_0(\gamma) \cos \omega_c t + J_1(\gamma) \cos (\omega_c t + \omega_m t) - J_1(\gamma) \cos (\omega_c t - \omega_m t) \quad (4.17) \\
&\quad + J_2(\gamma) \cos (\omega_c t + 2\omega_m t) + J_2(\gamma) \cos (\omega_c t - 2\omega_m t) \\
&\quad + J_3(\gamma) \cos (\omega_c t + 3\omega_m t) - J_3(\gamma) \cos (\omega_c t - 3\omega_m t) \\
&\quad + J_4(\gamma) \cos (\omega_c t + 4\omega_m t) + J_4(\gamma) \cos (\omega_c t - 4\omega_m t) + \cdots].
\end{aligned}
$$

Equation (4.16) indicates that the phase modulation generates a series of sidebands at frequencies $n\omega_m$ (n: integer) away from the carrier frequency. The magnitude of the higher order sidebands increases with the modulation depth γ in accordance with the increase of $J_n(\gamma)$.

Figure 4.10 shows sample phase-modulated signals in the time and frequency domains for two different modulation depths. The time-domain signals barely indicate the difference in the modulation depth. The frequency-domain signals, on the other hand, show the effect of modulation depth clearly. When the modulation depth (or index γ) is 0.2, the sideband peaks are barely seen at $\pm$ 5 Hz away from the carrier frequency 20 Hz. Higher order sidebands are not visible. When the modulation depth is increased to 1, the height of the carrier frequency peak decreases by approximately 20% (from 5×10^4 to 4×10^4) indicating that the energy in the carrier frequency is transferred to the sideband frequencies. In addition, the higher order frequency peaks appear on both sides of the central (carrier) frequency. This is in contrast to the amplitude modulation case where the increase in modulation depth does not generate higher order frequency peaks.

4.3.3 Frequency modulation

Frequency modulation resembles phase modulation. Consider that the frequency changes as a function of time (called the instantaneous frequency f_i) in the following fashion

$$f_i(t) = f_c + B\cos(2\pi f_m t).\tag{4.18}$$

The phase of the wave then becomes as follows

$$\phi(t) = 2\pi f_c t + \int_0^\tau B\cos(2\pi f_m \tau d\tau) = 2\pi f_c t + \frac{B}{2\pi f_m}\sin(2\pi f_m t).\tag{4.19}$$

By putting $\beta = B/(2\pi f_m)$, we can write equation (4.19) as $\phi(t) = 2\pi f_c t + \beta \sin(2\pi f_m t)$, hence the frequency modulated wave as follows

$$\psi(t) = A\cos(2\pi f_c t + \beta \sin(2\pi f_m t)).\tag{4.20}$$

Equation (4.20) indicates that a frequency modulated signal has the same form as a phase modulated signal. However, from the viewpoint of application, the frequency modulation is different from the phase modulation. It modulates the frequency itself so that the sensor feels a Doppler shift. This fact is applied to the technique known as laser Doppler vibrometry (LDV) [8]. This technique is widely used to measure vibration of an object applying frequency modulated laser light to the object surface and analyzing the reflected signal that contains the vibration of the object surface. The LDV detects the vibration of the object with optical interferometry using the laser light as the reference signal. The frequency modulation is necessary to make the interferometer sensitive to the direction of object vibration. The interferometer is sensitive to the phase change but not the direction of the object vibration without the frequency modulation.

References

[1] Yariv A 1971 *Introduction to Optical Electronics* (New York: Holt, Rinehart and Winston) ch 6

[2] Siegman A E 1986 Lasers (Sausalito, CA: University Science Books)

[3] Adhikari R, Bengston A, Buchler Y, Delker T, Reitze D, Shu Q, Tanner D and Yoshida S 1998 Input Optics Final Design, LIGO-T980009-01-D

[4] The Virgo collaboration 2006 The Virgo Physics Book, vol II: OPTICS and related TOPICS, April 21

[5] Miyasaka C 2009 Acoustic microscopy applied to nanostructured thin film systems *Modern Aspects of Electrochemistry* vol 44 (New York: Springer) pp 409–50

[6] Yamamoto H, Barton M, Bhawal B, Evans M and Yoshida S 2006 Simulation tools for future interferometers *J. Phys.: Conf. Ser.* **32** 398–403

[7] Yoshida S 2010 National Science Foundation Annual Report: 0653233 for Period:07/2009–06/2010

[8] OFV-5000 Vibrometer Controller. Available online: http://www.polytec.com/us/products/vibrationsensors/single-point-vibrometers/modular-systems/ofv-5000-vibrometer-controller/ (accessed on August 10, 2017)